T0203540

Constructing Low-rise Confined Masonry Buildings

Praise for this book

'Constructing Low-rise Confined Masonry Buildings is a very much-needed publication to promulgate confined masonry as a far safer alternative to both unreinforced masonry and RC frame and masonry infill construction – especially in earthquake-prone areas. It is a practical construction guide that communicates clearly and simply. Further, it is comprehensive in its coverage, and shows readers in a step-by-step sequence how to build not only confined masonry walls, but all the other elements necessary for a confined masonry building. Very well illustrated with several hundred drawings and photographs, the book is ideal for masons, builders, construction supervisors and architects. It deserves wide circulation.'

Andrew Charleson, Associate Professor, School of Architecture,
Victoria University of Wellington

'This is a unique guide which illustrates the construction process of low-rise confined-masonry buildings, and provides invaluable information on good construction practices in a user-friendly, but also comprehensive manner. This publication will be a vital resource for a variety of users interested in promoting and practising confined-masonry construction in earthquake-prone regions of the world.'

Dr Svetlana Brzev, Chair, Confined Masonry Network,
Earthquake Engineering Research Institute

Constructing Low-rise Confined Masonry Buildings

A guide for builders and architects

Tom Schacher and Tim Hart

PRACTICAL ACTION
Publishing

Practical Action Publishing Ltd
The Schumacher Centre, Bourton on Dunsmore, Rugby, Warwickshire,
CV23 9QZ, UK
www.practicalactionpublishing.org

First published by EERI 2015
This edition published by Practical Action Publishing 2018

A catalogue record for this book is available from the British Library.
A catalogue record for this book has been requested from the Library of Congress.

ISBN 978-185339-991-6 paperback
ISBN 978-185339-990-9 hardback
ISBN 978-178044-990-6 library pdf
ISBN 978-178044-991-3 epub

Citation: Schacher, T. and Hart, T. (2018) *Constructing Low-rise Confined
Masonry Buildings: A guide for builders and architects*, Oakland, USA,
Earthquake Engineering Research Institute, and Rugby, USA Practical Action
Publishing, <http://dx.doi.org/10.3362/9781780449906>

Since 1974, Practical Action Publishing has published and disseminated books
and information in support of international development work throughout
the world. Practical Action Publishing is a trading name of Practical Action
Publishing Ltd (Company Reg. No. 1159018), the wholly owned publishing
company of Practical Action. Practical Action Publishing trades only in support
of its parent charity objectives and any profits are covenanted back to Practical
Action (Charity Reg. No. 247257, Group VAT Registration No. 880 9924 76).

Cover photo: A mason at work in Ecuador
Cover design: RCO.design
Typeset by vPrompt eServices, India
Printed in the United Kingdom

http://dx.doi.org/10.3362/9781780449906.000

CONTENTS

ACKNOWLEDGMENTS

This guide has been prepared by Tom Schacher (primary author) and Tim Hart who are both members of EERI's Confined Masonry Network.

Particular thanks are due to Marjorie Greene without whose regular reminders this book might have remained at the 'intention-phase'.

This document grew out of workshops held at the Indian Institute of Technology in Kanpur in 2008 and the Pontificia Universidad Católica del Perú in 2009. Working groups were formed and the contents of various guides to be prepared were identified. A *Seismic Design Guide* was the first to be released on the EERI website in 2011. The *Engineered Construction Guidelines* is forthcoming. In 2013 a *Guidebook to Build Safer Houses in Confined Masonry* has been prepared by the humanitarian aid team of the Swiss Agency for Development and Cooperation in Haiti. Visit www.confinedmasonry.org to view these publications and others.

The authors gratefully acknowledge the comments and suggestions made by the contributors and reviewers: Andrew Charleson, Editor-in-Chief of the World Housing Encyclopedia (New Zealand); Svetlana Brzev, British Columbia Institute of Technology (Canada); Roberto Meli and Juan José Perez Gavilén, Universidad Nacional Autónoma de México; Leonardo Flores, CENAPRED, Mexico; Marcial Blondet, Pontificia Universidad Católica del Perú; Ahmed Mebarki, MSME Université de Paris-Est (France); Qaisar Ali, University of Engineering and Technology Peshawar (Pakistan); Junwu Dai, China Earthquake Administration, Harbin (China); Francisco Crisafulli, Universidad Nacional de Cuyo, Mendoza (Argentina).

Special thanks of course to all the people who have prepared illustrations and provided photographs. In addition to Tom Schacher, they are: Mariana Chavez (France), Nadia Carlevaro (Switzerland), Omid Esmaili (U.S.A.), Dorothea Haznas (Romania), Guillaume Roux-Fouillet (France/Switzerland), Victor Sanjines (Peru), Martin Siegrist (Switzerland), and Maggie Stephenson (Ireland).

Thanks also to Nadia Carlevaro (Switzerland) for her valuable feedback from the field training in Haiti; to Elizabeth Hausler Strand and the staff of Build Change for their contributions to the quality control recommendations and feedback; and to all the EERI volunteers and interns who have worked on the various draft versions of the manual.

The authors would also like to thank those organizations that have provided financial support for the creation of this manual: the Swiss Reinsurance Company (Zurich), the Swiss Agency for Development and Cooperation (SDC) for its own contribution and for allowing us to use the award fund from the Holcim Awards Acknowledgement Prize 2008 for the Asia Pacific region, the International Committee of the Red Cross (Geneva) and the Swiss Solidarity fund-raising organisation (Geneva). The authors would also like to acknowledge the financial support of Risk Management Solutions at the initiation of the project.

Most photographs and illustrations have been prepared by Tom Schacher.

The following illustrations have been supplied courtesy of Marcial Blondet and have been taken from his manual "Construcción y manteniemiento de viviendas de albañilería" (Lima, 2005): Fig. 9, 12 (left), 16, 17, 19, 22, 32a)+b), 33, 35, 36, 87, 90, 92, 105b, 106, 131, 137, 140, 145, 161 (left), 162 (all 3), 236.

Photograph No. 9 on page 23 supplied courtesy of Maggie Stephenson.

PREFACE

In most countries modern low-rise residential construction is made either of unreinforced masonry or reinforced concrete moment frames with masonry infill walls. Experience has shown that both systems can be signficantly affected by earthquakes. Unreinforced masonry buildings cannot deal adequately with horizontal forces, while reinforced concrete frames are difficult to build correctly. Too many details have to be observed and properly implemented, a challenge beyond the capabilities of self-builders or workers with no formal training. A simpler and more forgiving construction technology is needed to ensure safe construction.

Confined masonry combines elements of both systems, but it is a simple and forgiving construction method which has demonstrated good performance in past damaging earthquakes in Latin America. It is widely practised in Latin America and Mediterranean Europe, it has been the subject of laboratory testing and research studies, and has been incorporated in national building codes. In some Asian countries, such as Indonesia and China, confined masonry is regarded as a standard construction technique, while on the Indian subcontinent confined masonry has been actively promoted over the last decade.

As confined masonry is a construction system that has been developed by practitioners in various countries in parallel, there is a lack of uniform rules on how it should be implemented correctly. In 2008 the Confined Masonry Network decided to tackle this issue by compiling a set of common rules from the various existing codes and guidelines on confined masonry and use them to develop a uniform set of guidelines.

A first result is the *Seismic Design Guide for Low-Rise Confined Masonry Buildings,* published by the Network in 2011 (www.confinedmasonry.org), which provides prescriptive design provisions for engineers who want to use this construction system.

The present construction guide addresses the needs of small-scale contractors, technicians, government staff, architects as well as non-governmental organizations involved in post-disaster reconstruction. The guide has been written with users with various professional backgrounds in mind, including a workforce with little formal training. As a consequence this guide not only shows the practical detailing of confined masonry construction, but also offers a wealth of basic information on good construction practices in general.

A note of caution: this guide explains how to build **earthquake resistant houses with a maximum height of two stories (ground floor and upper floor)**. For taller buildings an experienced engineer must be consulted for specific calculations.

About the authors

Tom Schacher is an architect with extensive experience in humanitarian aid projects (Kenya, Rwanda, Turkey, Ethiopia) and with training in earthquake resistant construction techniques in Kenya, Rwanda, Turkey, Ethiopia, Iran, Pakistan, Haiti and Ecuador for the last 20 years.

Tim Hart is a civil and structural engineer at the Lawrence Berkeley National Laboratory, and has over 20 years' experience designing confined masonry buildings in Indonesia, China, Haiti, The Philippines, and Nepal.

Part I:
General aspects of confined masonry construction

http://dx.doi.org/10.3362/9781780449906.001

CHAPTER 1   WHY USE CONFINED MASONRY?

Confined masonry construction has been used during the last half century in various parts of the world. Researchers in Latin American and European countries have studied its behaviour and refined the technique, and governments have promoted its application with very satisfactory results. The severe 2010 earthquake in Chile (M 8.8) caused a relatively low number of victims, in part due to the wide use of confined masonry construction for single family and apartment housing.

Before going into the details of confined masonry construction, it is worthwhile to look at the two of the most common construction methods used for low-rise housing. They are unreinforced masonry and reinforced concrete frames with infill walls. It is important to understand their weak points which can be avoided by employing the confined masonry construction technology.

1.1   Unreinforced masonry in earthquake prone areas

Unreinforced masonry works well in areas with no earthquakes because masonry is strong in compression and the walls only have to bear vertical loads. In addition, masonry construction provides thermal comfort and durability. However, in earthquake-prone areas, horizontal loads due to earthquake shaking must be taken into account. Unreinforced masonry walls have some strength to resist lateral forces. However, this is limited and once it is exceeded the walls degrade rapidly, never to recover, like a stack of books on a wobbly table (Figure 1).

Figure 1: Stack of books not attached together will fall apart when the table gets shaken

In a building where walls are not well-connected to the floors and roof by reinforcing or confining elements, the walls will separate at the corners and the structure will undergo serious damage or collapse.

When unreinforced masonry is confined by concrete tie elements, however, it greatly slows down the degradation of the walls. Similar to a pile of books held together by a string, the books will still slide around but the string prevents the stack from falling apart (Figure 2). That's exactly what confined masonry is doing by holding all elements together with reinforced concrete ties (Figure 3).

3

Figure 2 and 3: Reinforced concrete ties hold the house together like a string around a stack of books

1.2 Reinforced concrete frames with infill walls in earthquake prone areas

In confined masonry the masonry walls are built first, and then concrete vertical and horizontal ties are poured around them. By contrast, in framed infill construction the concrete is poured first, then the masonry infill is placed (Figure 4).

Confined masonry: walls first, concrete ties later

RC frames with infill walls: concrete frame first, walls later

Figure 4: Inverse construction sequence in confined masonry compared to RC frames with infill walls

Reinforced concrete (RC) frame structures with infill walls are much more complex to build than they appear to the common worker. Concrete is not simply 'a glue' that holds everything together. Instead, the concrete frame is the primary force-resisting element. The masonry is placed after the concrete and is assumed to act as a non-load bearing partition. Because the concrete is the seismic load resisting element, special detailing is needed in the frame. This special detailing makes reinforced concrete frames very sensitive to implementation errors which can be fatal under earthquake conditions. A whole series of steps must be made properly:

- The rebar has to be placed in exactly the right position
- There must be enough space around the rebars to allow for sufficient concrete cover
- Rebar connection details must be correct
- Concrete has to be made in the right proportions and has to be mixed perfectly

- Pouring of concrete and its compaction must be done to a high standard (very difficult to achieve without a needle vibrator)
- Curing must be done properly to ensure correct hardening of the concrete.

The stress in the concrete moment frame members is high because the infill masonry is assumed not to contribute any strength to the wall. All vertical and horizontal loads have to go through the frame and its relatively narrow joints. Thus, there is no space for error. See Murty et. al. (2006) for a more detailed discussion of issues with RC frames.

The second problem with reinforced concrete frames is that the frames in themselves do not complete a building. A house needs walls. Yet, by putting walls in between the reinforced concrete columns, it doesn't allow for the frames to move freely as is often assumed by many engineers. This problem can be solved with special detailing, but the solutions may be too complex to be used by simple builders in poor countries.

Finally, infill walls are very difficult to hold back against out-of-plane loads (earthquake acceleration in the direction perpendicular to the walls). Infill walls can fall out of frames and can endanger people's safety.

Finished RC frames with infill walls and confined masonry buildings look very similar. However, these two systems have very different structural behaviour when subjected to lateral loads. In a RC frame system the load path is complex and all forces have to be carried by beams and columns and beam-column joints, while the load path in a confined masonry system is simple and straightforward because the forces are carried to the ground by the walls; it is a load-bearing wall system.

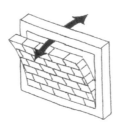

Figure 5: Walls tend to fall out of their RC frames

1.3 Confined masonry basics

In the confined masonry system the walls carry all the vertical and horizontal loads. Walls are held together by reinforced concrete confining elements that improve the in-plane strength of these masonry walls, resisting the shear forces induced by an earthquake (Figure 6).

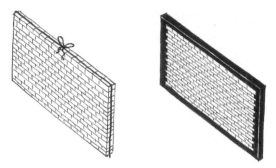

Figure 6: Confined elements called ties hold the walls together like a string

While detailing of rebar connections must still be correct, in particular the lap lengths, confined masonry structures are more tolerant of bad execution (Figure 7). If the concrete in the reinforced concrete ties is less than perfect, the system still works. However, the better the execution, the stronger the building.

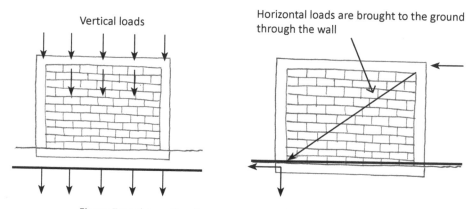

Figure 7: In the confined masonry system, load paths are simple

http://dx.doi.org/10.3362/9781780449906.002

CHAPTER 2   GENERAL RULES FOR BUILDING WITH CONFINED MASONRY

As with any other system, a certain number of rules have to be followed when a confined masonry building is constructed. Some rules are common sense or simply good building practice, while others are specific to confined masonry. These basic rules will be presented in the following sections. More in-depth technical information and construction details will be presented in Part Two of this guide.

2.1   Choosing a safe building site

It is important to select a construction site that is safe from natural hazards. There is no point in building an earthquake-resistant house if it is placed in a dangerous location (Figures 8-14). In densely populated urban areas suitable building sites are limited, particularly for the poorest sectors of the population. Architects, technicians and planners are responsible for informing house owners about the risks they are facing when building on a dangerous site, and to help them to find better locations, or at least assist them in reducing vulnerability. Safe locations are those where neither torrential rains nor secondary earthquake effects can damage a building. Building on steep slopes should be avoided because rain or earthquakes can trigger landslides or loosen boulders (Figure 10).

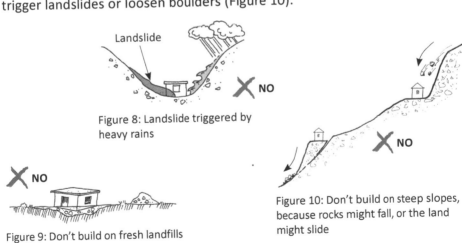

Figure 8: Landslide triggered by heavy rains

Figure 9: Don't build on fresh landfills

Figure 10: Don't build on steep slopes, because rocks might fall, or the land might slide

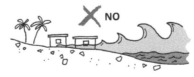

Figure 11: Don't build too near the sea: storms or tsunamis can flood the area

Figure 12: Don't build near river beds or at the bottom of a canyon or narrow valley

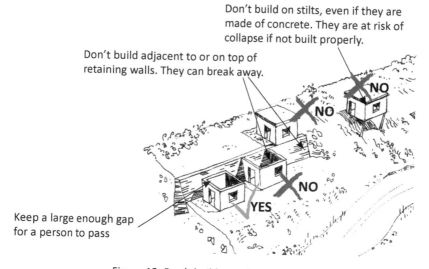

Figure 13: Don't build on stilts or adjacent to retaining walls

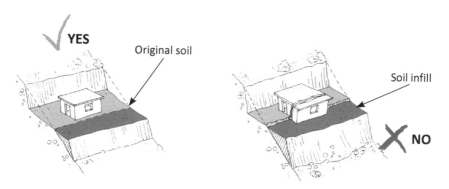

Figure 14: Don't build on top of fresh infills--try to build on original soil.

2.2 Main components of a confined masonry building

The shape of a building plan and the structural elements of a building influence earthquake performance. The following rules in Figure 15 apply to most construction technologies and building sizes.

Slab:

Slabs distribute earthquake loads to all the walls. They have to be well connected with the walls by means of horizontal and vertical ties.

Horizontal and vertical ties:

The vertical ties ("tie-columns") are the vertical confining concrete elements of the walls. The horizontal ties (plinths and ring beams) are the confining concrete elements above and below the walls. All ties must be securely fastened to one another.

Solid walls:

Each façade should contain at least one solid wall panel without openings.

Openings:

All openings must be confined. Openings should not be too large and should be well distributed.

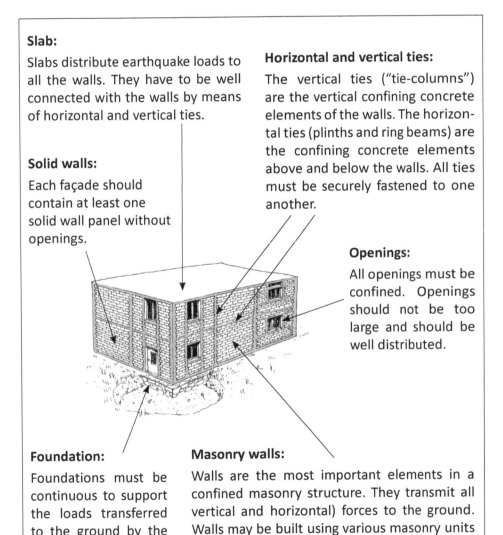

Foundation:

Foundations must be continuous to support the loads transferred to the ground by the walls.

Masonry walls:

Walls are the most important elements in a confined masonry structure. They transmit all vertical and horizontal) forces to the ground. Walls may be built using various masonry units of adequate quality and strength. Walls must be well distributed though the building plan.

Figure 15: Main elements of a confined masonry building

2.3   Building configuration

Height:

While some of the illustrations might show taller buildings, the rules and guidelines presented in this guide apply to one- and two-storey buildings, with a ground floor plus one floor above (Figure 16).

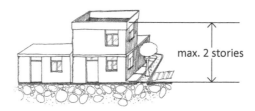

Figure 16: Maximum height of buildings using the rules of this guide

Regularity:

Simple shapes and regular layouts are very important factors in designing earthquake resistant buildings (Figures 17 and 18). Symmetrical wall layouts are equally important and are dealt with in the following section.

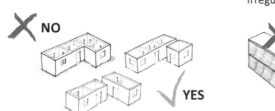

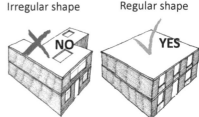

Figure 18: Buildings with complex forms should be divided into several independent simple forms

Figure 17: A building should have a regular shape

Plan length to width:

The building plan should be a regular shape. The building should not be excessively long. Ideally, the length-to-width ratio in plan should not exceed 4 (Figure 19).

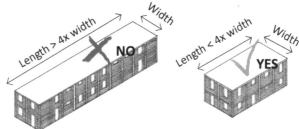

Figure 19: A building must not be longer than 4 times its width

Distance between buildings (seismic gap):

Buildings too close to one another may hammer against each other during a quake, resulting in damage. This phenomenon is known as pounding. If buildings have different heights, they might even pose a higher risk, pounding at particularly sensitive points and thus accelerating collapse.

The theoretical minimum distance between buildings would be the sum of expected displacements at the roof level, however, in reality such a narrow gap is hard to execute properly because no mortar or debris must fall into the gap, blocking free movement. It is therefore suggested to keep a minimum gap of 10 cm (4 in.) between two-storey buildings. A bigger gap (45 cm to 60 cm) is preferable as it allows a person to pass through (Figure 20).

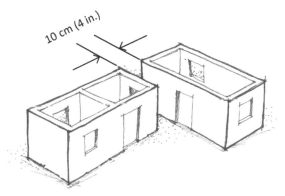

Figure 20: Minimum gap between buildings: 10 cm (4 in.)

2.4 Walls

The essence of the confined masonry system lies in the strength of its walls which bear all vertical (gravity) and horizontal (earthquake) loads. For the walls to be able to do that, they need to be confined with reinforced concrete elements, must not have any openings and should be at least as long as they are high. It is these shear walls which will ensure the earthquake resistance of a building (Figure 21).

A sufficient number of confined masonry walls without openings must be placed in each direction and should be distributed as evenly as possible over the floor layout. For how to calculate the exact number of walls, please see the appendices.

- Each facade needs at least one solid wall panel
- A confined wall panel can only be used as a shear wall if its length is more than two thirds of its height.

11

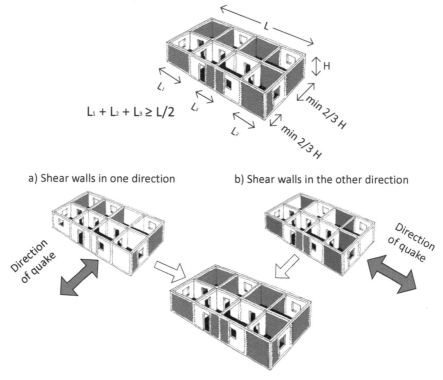

$$L_1 + L_2 + L_3 \geq L/2$$

a) Shear walls in one direction   b) Shear walls in the other direction

Figure 21: Amount, proportions and position of shear walls

Wall layout:

Symmetrical wall layouts are the best in order to achieve the same number of walls on either side of the building. Keeping all small rooms (with a lot of walls) on one side and large rooms (with few walls) on the other side leads to an irregular stiffness of the building. Uneven distribution of walls can cause a building to twist (torsion) during a quake, leading to collapse (Figure 22).

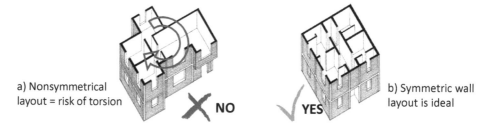

a) Nonsymmetrical layout = risk of torsion   **NO**   **YES**   b) Symmetric wall layout is ideal

Figure 22: Irregular distribution of walls leads to torsion

Vertical continuity:

Always build second floor walls directly over first floor walls. Don't build overhanging floors (Figure 23).

No vertical wall continuity
Overhanging walls are not permitted

Walls at upper floor walls must be
placed over lower walls

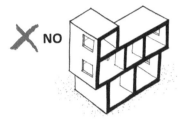
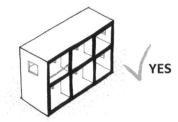

Figure 23: Upper walls must be placed on lower walls (continuity of load paths)

Never place an upper floor on a series of ground floor columns. Lower floors should be stiffer than or equally stiff as upper floors. To achieve this, as many walls as possible are needed at the lower floor, in both directions (Figures 24 and 25).

Figure 24: This portico will cause the building to rotate during a quake

Figure 25: Never place a building on columns

Wall lengths:

Confined wall panels between adjacent transverse walls should not be longer than 4.5 m in areas of high seismicity, and 6 m in areas of moderate seismicity (see Meli et al., 2011, for more discussion) (Figure 26). This is due to a risk of out-of-plane wall damage. In buildings with single brick walls of less than 15 cm (6 in.) the maximum length of walls between confining tie columns should be reduced to 4 m. (Figure 27).

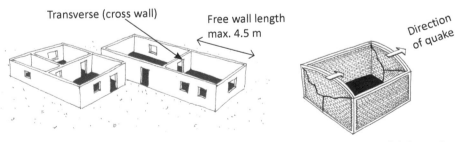

Transverse (cross wall)

Free wall length max. 4.5 m

Direction of quake

Figure 26: Maximum wall length

Figure 27: Out-of-plane wall deformation

2.5   Foundation

A frequent error observed at building sites is the use of isolated footings under the 'columns'. This mistake can be a consequence of the erroneous use of the word 'column' and the inherent lack of understanding of the structural system of confined masonry as well as a perceived need to economize on building material.

Columns and isolated footings belong to the world of RC frame structures where columns carry the weight of the building. In the confined masonry system the walls carry most of the weight. The foundation therefore has to be continuous (Figure 28).

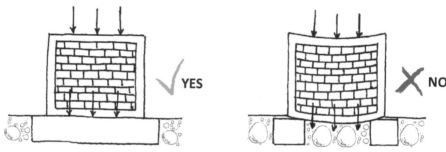

Figure 28 a): Continuous foundations transmit the load correctly to the ground

Figure 28 b): Isolated footings under a load bearing wall cannot transmit the load correctly

2.6   Openings

Windows and doors are essential elements in a building, but unfortunately they do weaken confined masonry structures. From a structural point of view, the fewer openings the better (for architects it is usually the other way around!). If the opening has an area greater than 10% of the wall panel area, it cannot be considered a shear wall. A wall that has a small opening with an area less than or equal to 10% of the wall panel area could be considered a shear wall depending on where the opening is placed and how it is confined (see Meli et al. 2011 for more discussion of how to treat small openings).

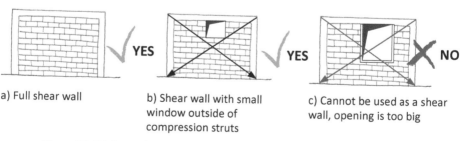

a) Full shear wall

b) Shear wall with small window outside of compression struts

c) Cannot be used as a shear wall, opening is too big

Figure 29: Wall panels with big openings cannot be calculated as shear walls

Size:

In walls that are not considered as shear walls, the width of openings should not exceed half the length of the wall panel because they still do support vertical loads.

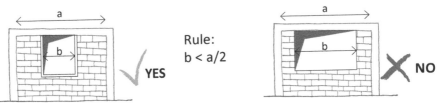

Rule:

b < a/2

a) Correct: window is smaller than half the length of the wall panel

b) Incorrect: window is larger than half the length of the wall panel

Figure 30 a) and b): Wall openings must be smaller than half the length of the wall panel

Confinement:

All openings must be confined by reinforced concrete ties. Such ties may be smaller than the main vertical or horizontal ties, with only two instead of four lengthwise rebars. Vertical ties, spanning between the plinth and the ring beam ties, are more effective, but also more complicated to put in place. Horizontal ties (seismic bands) are somewhat less effective (the distance between the vertical main ties being greater than the distance between the horizontal main ties) but much easier to put in place (Figure 31).

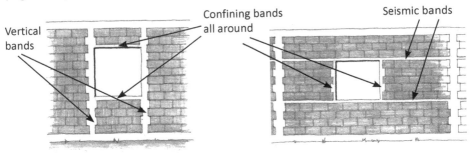

a): Confinement of opening tied vertically to the main ties: more effective, but more cumbersome

b): Confinement of opening tied horizontally to the main vertical ties: easier to put in place

Figure 31 a) and b): All openings must be confined and tied to the main vertical or horizontal ties

Distribution:

Openings should be placed in the same position at each floor level (Figure 32). The window area should be reduced to a minimum. Care should be taken when placing an opening close to corners, as this may weaken the structure.

 YES

 ✗ NO

Figure 32 a): Windows should be placed in the same position on each floor

Figure 32 b): The number of openings should be reduced to a minimum and be vertically aligned

2.7 Slabs

Floor slabs are important elements, particularly in buildings higher than one storey. They act as horizontal diaphragms; that is, they ensure that horizontal forces are equally distributed among all load-bearing walls. Floor slabs can be made in the form of concrete slabs or wooden floors. They should have a simple layout and as few openings as possible (Figure 33).

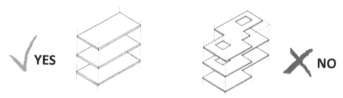

Figure 33: Floor slabs should have a simple shape and few openings

2.8 Roofs

Single storey houses, or the top floors in multi-storey buildings, may also be covered by light framed roofs rather than by concrete slabs. People who have lived through a damaging earthquake are often uncomfortable with the idea of sleeping under a concrete slab. Light framed roofs do not provide the same kind of rigidity as a slab does. However, by observing some basic rules, roofs made with timber or steel structures can give satisfying results.

There are two basic types of roof shapes: two-sloped pitched roofs and four-sloped hipped roofs. A pitched roof requires gable walls which have proven to be vulnerable in earthquakes. If masonry gable walls are not properly reinforced (confined) they tend to suffer severe damage, leading to the collapse of the roof (Figure 34).

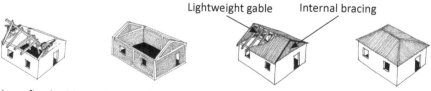

Unconfined gable walls may collapse, while confined walls won't

Lightweight gable with internal wind braces

Hipped roofs can work without internal bracings

Figure 34: Confined gable walls and bracing of the roofs

16

Gable walls can also be made out of timber, making them lighter and less prone to collapse. However, in this case internal bracing of the roof structure is paramount. Hipped roofs do not have this problem, as the hip rafters act as a bracing.

2.9 Summary of do's and don'ts related to confined masonry construction

The following two illustrations (Figures 35 and 36) present a summary of the rules explained before:

A vulnerable house

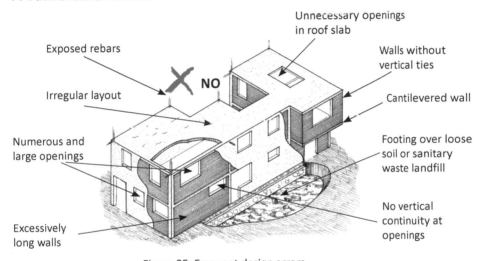

Figure 35: Frequent design errors

An earthquake resistant house

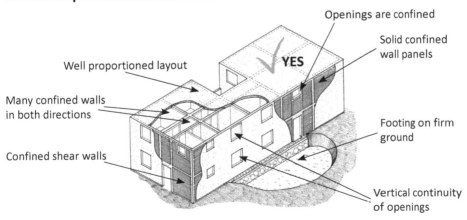

Figure 36: A house made following the rules

2.10 Building materials

Because in the confined masonry system walls do most of the job of bringing the vertical and horizontal loads down to the ground, the quality of the masonry units employed is extremely important. The compressive strength of masonry units should be above 10 MPa.

Burnt clay bricks:

As a general rule, solid bricks are preferred to multi-perforated ones. Multi-perforated clay brick with vertical holes covering less than 50% of their horizontal surface are also admissible (Figure 37).

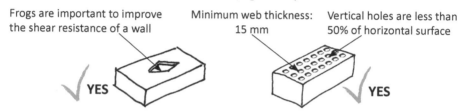

Frogs are important to improve the shear resistance of a wall

Minimum web thickness: 15 mm

Vertical holes are less than 50% of horizontal surface

Figure 37: Solid burnt clay bricks and burnt clay bricks with less than 50% of vertical holes are both admissible

Bricks with bigger or with horizontal holes are not admissible. They are too weak to be used in a load-bearing function (Figure 38).

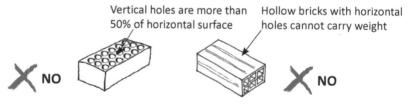

Vertical holes are more than 50% of horizontal surface

Hollow bricks with horizontal holes cannot carry weight

Figure 38: Bricks with more holes than 50% of their surface or with horizontal holes must not be used.

Quality check for burnt clay bricks:
Burnt clay bricks can come from a variety of sources ranging from bricks made in local makeshift kilns to high quality bricks from industrial plants. Before buying them, it is wise to check their quality with some simple tests. Only quality bricks with a minimum resistance to compression of 10 MPa (100 kg/cm^2) should be used.

Visual check:

Bricks must:

*   be regular in form
*   be uniform in colour
*   not be warped
*   not show any flaws or lumps

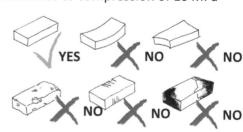

Figure 39: Defects in bricks

Physical check:

- Bricks must not be easily scored by a knife.
- Bricks must pass the 3 point test (man standing on a brick spanning between two other bricks.
- Bricks must give a ringing sound when struck against each other.

Figure 40: Brick test: resist 3 point test and give a ringing sound when struck together

Concrete blocks:

Concrete blocks, also called cinder blocks or cement blocks, are very popular all over the world and are particularly appreciated in poor countries with no brick industry. They are quite cheap to produce, making them the masonry unit of choice for self-builders.

However, concrete blocks may be weak. With few exceptions only the blocks from industrial plants reach the required compressive strength of at least 10 MPa (in industrialized countries, cement blocks have compressive strengths of up to 40 MPa).

Blocks made by local small-scale producers have compressive strengths which frequently do not exceed 4 MPa. There are a number of reasons why quality is often very low and why it is difficult for small-scale producers to make good blocks:

- Small-scale producers work in tiny yards. They have to sell their blocks the day after production to be able to start with a new batch. Blocks are dried in the sun to make them look well-finished, but without the necessary 10-day curing they don't harden, even if they dry.
- Economic limitations do not allow for the acquisition of a vibrating press, tarpaulins and a sufficient amount of water for curing.
- To lower the cost, the cheapest (and unsuitable) aggregates are used, and the quantity of cement is reduced to a minimum.
- The producer's lack of technical knowledge hinders a correct choice of aggregates, a correct mixing of the ingredients or an understanding of the importance of curing.

Having said that, there are small-scale producers who make good blocks! Thus, before buying concrete blocks one is well advised to check on the producers capacity. Their low quality can cancel out all efforts put into correct construction details and execution. Testing the blocks on a regular basis, ideally through a recognized laboratory, is highly recommended (Figure 41).

The choice of blocks:

Concrete blocks must be strong. Solid concrete blocks are excellent, though hard to find. Hollow blocks with three lengthwise walls are ideal to transmit in-plane forces. Blocks with only two lengthwise walls but three or four cross-walls are acceptable if of good quality.

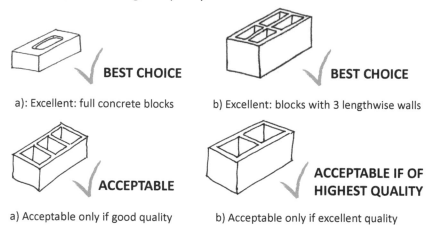

BEST CHOICE

a): Excellent: full concrete blocks

BEST CHOICE

b) Excellent: blocks with 3 lengthwise walls

ACCEPTABLE

a) Acceptable only if good quality

ACCEPTABLE IF OF HIGHEST QUALITY

b) Acceptable only if excellent quality

Figure 41: Recommended types of blocks

Block dimensions:

Blocks should have a minimum width of 15 cm (6 in.). However, it is preferable to use 18 cm or 20 cm blocks (Figure 42).

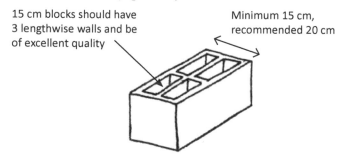

15 cm blocks should have 3 lengthwise walls and be of excellent quality

Minimum 15 cm, recommended 20 cm

Figure 42: Minimum width of blocks is 15 cm (6 in.). A width of 20 cm (8 in.) is preferable

In hollow blocks the voids must be less than 50% of the total block surface. External walls must not be less than 2 cm (7/8 in.) thick (Figure 43).

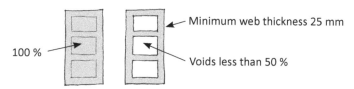

Minimum web thickness 25 mm

100 %

Voids less than 50 %

Figure 43: Net surface (load bearing surface) of the blocks must be above 50 %

Site selection:

1. Don't build on a steep slope or on stilts
2. Don't build near rivers that can flood

Site selection:

3. Don't build on top of unstable retaining walls
4. Don't build next to river banks which can erode
5. Don't build on top of cliffs; they can slide away

Design principles

6. Don't make too many windows
7. Build shear walls on each façade
8. Gable walls fall out if not confined
9. Don't build overhangs

Design principles

10.   Don't build overhangs

11.- 14.   Too many and too large windows and no shear walls weaken the building

Quality check for concrete blocks:
For the reasons explained above, concrete blocks must be tested before use. The following checks and tests give a first impression, but do not replace a proper lab test.

Visual check:
- Blocks must be regular in shape, with the top and bottom surfaces parallel.
- Compare various blocks from the same producer to make sure that they have the same height (caution: heights can vary from one producer to another).
- The concrete of the blocks should have an even surface finish with no honey combing (gravel nests).
- Maximum size of aggregates used for the concrete must not be larger than 1/3 of the wall thickness.
- It must not be possible to break off parts of the walls with your bare hands.

Physical check:
A simple way to test the quality of concrete blocks is to let them drop from the height of a person's head (1.5 m or 5 ft.) onto a hard concrete surface. Blocks must be held with their holes facing downwards and they should land flat on their lower face. They should not land on their corners or edges, as they will break more easily.

If 4 out of 5 blocks are not seriously damaged by the fall, then the batch has a minimal acceptable level of quality (Figure 44).

Proper lab tests however give more reliable results. For any larger construction, such as community buildings or multi-storey housing, blocks should be tested by a lab before purchase.

Figure 44: Drop test for concrete blocks

Cement:

Portland cement is used for both concrete and mortar. It is a fine gray powder delivered in paper bags. With time, it will absorb moisture from the air and set in lumps. Cement which has lumps must not be used. Cement loses its binding quality over time:

Fresh cement:	100 % of binding capacity
After 3 months:	80 % of binding capacity
After 6 months:	70 % of binding capacity
After 1 year:	60 % of binding capacity
After 2 years:	50 % of binding capacity

Figure 45a: Effect of time on cement

Bags off the ground and away from walls

Storage:

Cement bags should be stored in a shed and away from the floor and walls to protect them against humidity (Figure 45b).

The bags should not be stacked higher than 4 layers as excessive compaction will encourage hardening. Also, they should be stacked close together to reduce the circulation of damp air.

Figure 45b: Keep cement in a dry shed

To ensure timely use of the cement, storage has to be organized on a 'first in - first out' basis.

Quality test:

Mix cement and water to form a small round cake. The hardening process has to start within 30 to 60 minutes. After 18 to 24 hours, the surface must be hard enough so it cannot be scratched easily with a finger nail (Figure 45c).

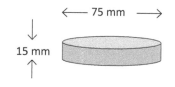

← 75 mm →

15 mm

Figure 45c: Test cake for cement freshness

Lime:

Lime is not used in the preparation of concrete, but is a very useful ingredient for good mortar. However it is often difficult to find in less industrialised countries. It makes the mortar less stiff and allows for smoother working. It is produced by burning limestone in a kiln. The resulting product is quicklime which is extremely aggressive to the skin and must not be touched. Quicklime is then transformed into two types of lime which can be used for specific purposes:

Hydrated lime
- Hydrated lime is quicklime 'slaked' with water. It comes in powder form or as lime putty (white tooth-paste like substance).
- It hardens in the presence of air (CO_2).
- Hydrated lime can be used for whitewash or plaster work.
- Pure hydrated lime should not be used together with cement in mortars as it may considerably slow down the hardening process (most hydrated limes however are more or less 'hydraulic').

Hydraulic lime
- Hydraulic lime is hydrated lime containing silica, alumina and iron oxides. It comes in a powder form and can be used in a similar way to cement.
- It hardens in the presence of water (thus 'hydraulic').
- Hydraulic lime, being similar in character to Portland cement, is an ideal addition to the latter. It increases the workability of the mortar and hardens at a similar rate.

'Masonry cement'
- Masonry cement is a ready-made mixture of Portland cement and hydraulic and hydrated lime.
- It can be used directly or be added to Portland cement to make a good masonry mortar.

Aggregates:
Sand and gravel are the aggregates used in concrete. They have to be of a variety of sizes so that the smaller grains can fill the gaps between the bigger grains. When the two components are mixed in the right balance, they are said to be well-graded. Because there are more particles touching each other in well-graded aggregates, the resulting concrete is stronger (Figures 46 and 47).

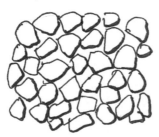 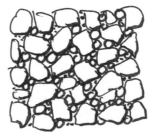

Figure 46: Gravel without sand has gaps between the grains

Figure 47: Sand particles of different sizes fill the gaps

Sand:

Sand is a major ingredient in concrete and the main ingredient in mortar. Round river sand and crushed quarry sand are equally good as long as they areclean and with no mud particles. Otherwise they need to be washed prior to use. Sea sand must not be used. It contains salt which is bad for the concrete and steel reinforcement bars.

The diameter of sand grains should vary. For concrete, a good mix of grains from 1 - 4 mm is fine. For mortar it is better to have a sand with even smaller grains. Though a high percentage of very fine particles increases the need for cement, the quality of mortar improves. The best mix for a good mortar would be the following:

Grain size	Amount
0 - 0.5 mm	1 part
0.5 - 2 mm	1 part
2 - 4 mm	1 part

Figure 48: Sand grading for mortar

Sand quality test 1:

Take a handful of sand with a slightly moist hand. If the hand remains dirty, the sand contains too much organic material and must be washed before use (Figure 49).

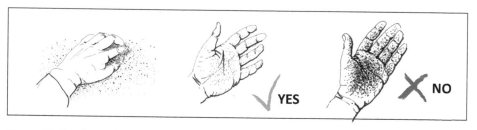

YES   NO

Figure 49: Simple sand test: if the hand remains clean, sand is OK, if it is dirty, sand must be washed before use

Sand quality test 2:

Put 15 cm (6 in.) of sand into a transparent water bottle. Add water to the top. Shake well and wait 3 hours to let the sand and fine particles settle. Measure the thickness of the layer of fine particle. It must not exceed 12 mm (1/2 in.) (Figure 50).

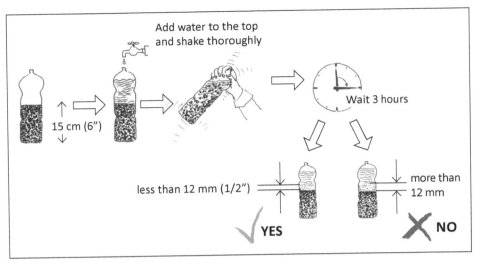

Figure 50: Proper sand quality test: fine particle layer must not exceed 8 % of solids

Gravel:

Gravel is the second aggregate component of concrete. It can be round or crushed. Crushed gravel makes a stronger concrete, but round gravel is more workable which is an important factor when filling in narrow columns or vertical ties and a needle vibrator (vibrating probe) is not available.

In order to ensure a strong concrete, gravel must come from hard rock and not from soft (lime) stone. In addition, gravel must be well-graded, with maximum grain size not exceeding 18 to 20 mm (3/4 in.).

Finally, gravel has to be clean and with no mud particles or it needs to be washed prior to use.

Cleanliness check of gravel:

Check the cleanliness by putting half a shovel of gravel into a bucket and fill up with water. Move the gravel for a while and check the shade of the water. If the water looks rather dirty, wash all gravel before use.

Water:

Water employed in construction must be clean, ideally of drinkable quality. Salty sea water must never be used. Wells near the sea shore must be checked regularly by tasting to make sure that the water is not salty.

Steel (rebars):

All rebars used in construction must be of the deformed type, except for the 6 mm (1/4 in.) stirrup bars which can be smooth (Figure 51). The steel should be ductile (i.e. ultimate elongation 9%).

Figure 51: Deformed and smooth rebars

Rebars have different (nominal yield) strengths. In confined masonry, grade 60 bars should be used (grade 60 = 60,000 psi = 400 MPa). In certain countries the steel grade is indicated on the rebar (Figure 52).

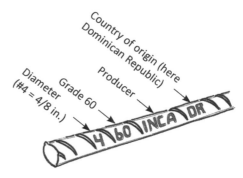

Figure 52: Indications on a rebar

Timber:

Timber is a precious and costly construction material.

- If timber is used and reused for temporary construction work (formwork, shoring), it must be cleaned and stored properly after each usage (Figure 53a).
- If timber is used for permanent features (e.g. roof carpentry) it must be treated against insects such as termites and well protected from rain and humidity coming out of the ground.

Timber for construction work must be dry and straight. Boards for formwork should not show cracks, splits or an excessive number of knots because the formwork must be watertight (Figure 53b).

Concrete needs water to harden. If water can seep out through cracks in the formwork, concrete does not only lose the moisture it needs for the chemical process of hardening, but seeping water will also carry away cement. This will result in gravel nests, insufficient concrete cover of rebars and generally weak concrete. (If it is impossible to find perfectly tight boards for formwork, the cracks can be filled with good clayey mud. However, this is a last resort solution which should be avoided if possible).

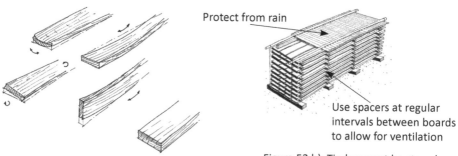

Figure 53 a): Defects in timber boards

Figure 53 b): Timber must be stored properly and protected from rain

2.11 Concrete and mortar

Concrete and mortar are conglomerates of rock fragments bound together by cement. Water allows the chemical reaction of the binding process.

Concrete used in construction is made of gravel, sand, Portland cement and water. Mortar is made of sand, Portland cement, lime (ideally) and water.

The right quantity of each ingredient (proportion), proper mixing, compaction and curing are important steps to ensure the required quality of the end product.

Concrete:

Mixing proportions:

The amount of cement contained in concrete has an important influence on its strength. Concrete exposed to environmental influences (rain, pollution, salt in marine environments) should have a minimum amount of 300 kg of cement per cubic metre of final fresh concrete. This corresponds to a concrete of class C25/30 (25 N/mm^2 for a test cylinder and 30 N/mm^2 for a test cube after 28 days).

In order to allow for a concrete not made properly, as often happens with an untrained workforce, the values taken into account in the present manual are lower. A compressive strength of a test cylinder of 15 N/mm^2 (15 MPa) has been assumed (concrete class C12/15). To achieve this compressive strength values, a minimum of 250 kg of cement per cubic metre of finished concrete is required (Figure 54).

The following table gives the equivalent mix proportions in volumes and cement proportions in kilograms per cubic metre of final concrete (kg/m^3):

	By volume Cement	By volume Sand	Gravel	=	Approximate ratio kg cement / m³ concrete
Minimum mix→	1	2	4	=	200 kg/m³
Standard mix →	1	2	3	=	250 kg/m³
Ideal mix →	1.5	2	3	=	350 kg/m³

Figure 54: Proportions of concrete ingredients

Example:

The mixing ratio of 1 : 2 : 3 is based on the following calculation:
- 250 litres of cement + 500 litres of sand + 750 litres of gravel. Sand and gravel combine to form 1 cubic metre of aggregates (sand filling a big part of the gaps between the gravel).
- Cement has an bulk density between 0.8 (loose) and 1.6 (highly pressed, as after a long truck journey). For measures based on shovel-filled buckets, a density of 1 seems to be appropriate.
- This means that 250 litres of cement weigh 250 kg, thus producing a mix of 250 kg/m³.

Water:

The amount of water added to the dry mix of cement and aggregates is of utmost importance. Too much water will cause the concrete to separate into its parts, and too little water will make it stiff and unable to make the necessary chemical reaction. Given however that usually there is already a certain amount of water in the aggregates, due to humidity or recent rains, it is quite difficult to give an exact amount of water to add. It is recommended to start with a quantity of water that corresponds to half of that of cement. If the mixture remains too dry (stiff) more water can be added. But, water should never be more than the volume of cement used in the mix (Figure 55).

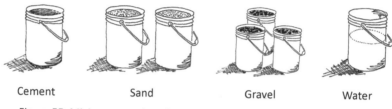

Cement   Sand   Gravel   Water

Figure 55: Mixing proportions for concrete: 1:2:3   +   0.5 to 1 of water

Mixing by hand:

Aggregates and cement must be mixed dry. Create a heap with gravel, sand and cement on top. Displace the whole heap with shovels onto a new heap. Shovel this second heap back again. At this point the ingredients should be thoroughly mixed.

Prepare a little 'lake' on top of the heap and fill it with water. Now shovel carefully the aggregate mix from the edges of the heap into the lake. Don't use too much water. Go on until the whole heap is thoroughly mixed, moist and with the right consistency (Figure 56).

Never add water to only a part of the heap, planning to mix it when you need it. The resulting concrete will be very inconsistent.

Figure 56: Hand mixing of concrete, from left to right: first dry mix twice, then add water

Mixing with a concrete mixer:

Mixing concrete with a mixer gives a much better and regular result (Figure 57). One good mixing method is the following:

1. Put half of the water and all the cement into the mixer

2. Turn on the mixer and let it mix for a minute

3. Add the aggregates

4. Let it turn for another minute

5. Add water slowly, bit by bit, until getting the right consistency

6. If the consistency is too liquid, add a little bit more cement

Figure 57: Concrete mixer

7. Don't let the mixing process last for more than 3 to 4 minutes.

In order to clean the mixer properly, fill it with one bucket of water and half a bucket of gravel. Let it turn for 10 to 15 minutes.

Quick test:

If the mix and the amount of water is correct, one must be able to press a handful of concrete into a small ball. If the concrete runs through the fingers, it is too liquid and should be corrected by adding more cement and some sand. If the aggregates remain separated and you're not able to form a ball, the concrete is too dry. More water can be added (Figure 58).

a) Concrete can be pressed into a ball, thus is correct

b) Concrete runs through the fingers, thus is too wet

Figure 58: Quick test of concrete consistency

Slump test:

The slump test shows the right consistency of concrete. A standard slump cone is filled in three equal layers with fresh concrete. Each layer is tamped down 25 times with a steel rod (ø 16 mm = 5/8"). Then the cone is lifted vertically and placed next to the slump. The difference between the slump and the cone should not be more than 12.5 cm (5 in.) (Figures 59 and 60).

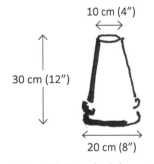

10 cm (4")

30 cm (12")

20 cm (8")

Figure 59: Standard slump cone

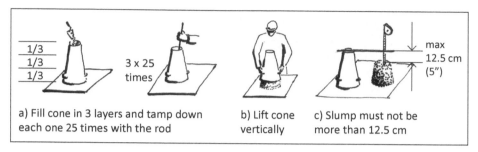

a) Fill cone in 3 layers and tamp down
each one 25 times with the rod

b) Lift cone
vertically

c) Slump must not be
more than 12.5 cm

Figure 60: Slump testing procedure

Working time:

Fresh concrete has to be used within 90 minutes. After this, concrete starts
to harden and must not be used any more. Never add water to refresh
concrete when it starts to harden. The quality of such refreshed concrete
is very bad.

Therefore, mix only the amount of concrete which can be used in one hour.

Don't pour concrete at zero degrees Celsius. It will freeze and not harden
properly.

Curing:

Concrete should not dry, but rather harden. In order to be able to do that,
concrete has to be *cured*. The aim of the curing process is to make sure that
the concrete can maintain its moisture so it can harden correctly.

Curing is done by spraying new concrete with water three times a day, for a
week, and covering it each time with a tarpaulin or empty bags of cement.
Spraying should start half a day after pouring the concrete, even if it is still
in the formwork (Figure 62). Slabs should be kept under water which can be
done by creating small ponds with sand barriers (Figure 61).

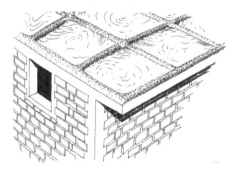

Figure 61: Curing a slab by keeping it under water

Figure 62: Curing newly poured
concrete by hosing it with water

Concrete hardening times:

Concrete gains its strength over time (Figure 63). For practical reasons concrete strength is generally measured after 28 days, at which point it usually has achieved 60 % of its final strength. The following table gives a more complete picture:

Strength of Portland Cement Concrete:		
3 days:	approximately	20 %
7 days:	"	45 %
28 days:	"	60 %
3 months:	"	85 %
6 months:	"	95 %
1 year:	"	100 %

Figure 63: Hardening times of concrete

By consequence, formwork can only be taken off when the concrete is reasonably hard. This will depend on the function and location of the concrete element. The following table gives some general indications about when the formwork can be removed (Figure 64).

Lateral parts of horizontal ties or beams:	2 days (but cover immediately with tarpaulins)
Lateral parts of vertical ties or columns:	3 days
Lower parts of slabs, beams or lintels • Span up to 3 m (10 ft.): • Span from 3 m to 6 m (10 to 20 ft.): • Span beyond 6 m (20 ft.):	8 - 14 days 16 - 24 days 24 - 35 days
Cantilevered elements:	As long as possible, but at least 35 days

Figure 64: Days to wait until removal of formwork

Mortar:

Mixing proportions:

Cement mortar made of sand and Portland cement is the most frequent type of mortar used in low income countries. This type of mortar should be made with 1 part of cement and 3 parts of sand. Mortar with a cement:sand ratio of 1:5 can be used only if the masonry blocks are greater than 15 cm wide. Ratios greater than 1:5 should not be used since it would make the mortar too weak.

Type	Cement	Sand	Compressive strength
Cement mortar	1 part	3 parts	12 N/mm²
Cement mortar	1 part	5 parts	4 N/mm²

Figure 65: Mixing rate of cement mortar

The ingredients and proportions of *cement - lime mortar* are a bit more complex as they depend on local traditions.

- In southern Europe (e.g. France, Italy and Spain) cement - lime mortar known as *bastard mortar* is made from Portland cement and hydraulic lime.

- In the Americas a similar mortar is known as *masonry mortar,* made with Portland cement and hydrated lime or mortar cement (itself already a mixture of cement and hydraulic and hydrated limes).

Cement - lime mortar has a lower compressive strength than simple cement mortar but offers a better workability and a higher elasticity which is ideal for confined masonry. However, in many countries lime (hydraulic or hydrated) can be difficult to come by. In this case pure cement mortar is the only alternative.

The mixing ratios of various mortars are given in the following table:

	Type	Cement	Hydraulic lime	Sand
Minimum mix ⟶	Cement mortar	1	2	9
	Cement mortar	1	1	6
Ideal mix ⟶	Cement mortar	1	0.5	4.5

Figure 66: Mixing rates of cement - lime mortars

Working time:

Fresh mortar should be used within one hour. Never add water to refresh hardening mortar. The quality of such refreshed mortar is very bad. Therefore, mix only the amount of mortar which can be used in one hour.

Don't use mortar at freezing temperatures. It will not be able to harden properly.

Curing:

As with concrete, mortar needs water to become hard. The curing process ensures that moisture remains in the mortar (Figure 67).

A first important step to ensure moisture in the mortar is to make sure that the masonry units (bricks, concrete blocks) will not suck the water out of the mortar. Thus, masonry units have to be drenched before use. This can be done by dunking the units into water, by applying water with a big brush or by spraying water over the units for a good amount of time.

PREFERRED METHOD   **ACCEPTABLE METHOD**

Figure 67: Drenching masonry units before use is very important

The proper curing process however starts once the walls are made: they have to be watered down regularly during the first week, ideally 3 times a day if the weather is hot. It is recommended to cover the walls with tarpaulins to make sure that the bed joints remain moist (Figure 68).

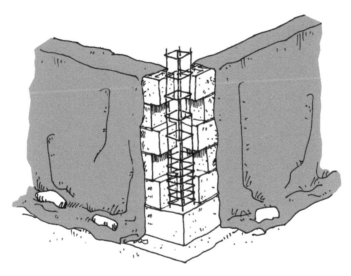

Figure 68: Cover the moist walls with plastic sheeting to avoid evaporation

Final check:

If it is possible to scratch off mortar in the joints easily after two weeks, the mortar is weak and has not hardened properly. In this case it might be necessary to demolish the wall and start all over again.

2.12 Tools

The tools shown here are the usual ones masons are using for modern construction work where they work with concrete and masonry (Figure 69).

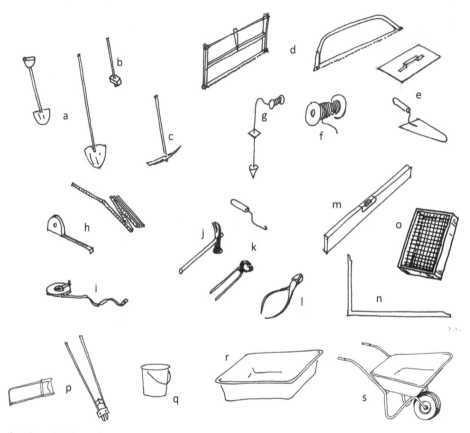

a) Shovel and spade

b) Sledge hammer

c) Pick axe

d) Saws

e) Trowel and float

f) String

g) Plumb line

h) Measuring tape or foldable meter

i) Long measuring tape

j) Claw hammer

k) Pincers or wire twister

l) Wire cutter

m) Spirit level

n) Mason's square

o) Sieve for sand

p) Metal saw or rebar cutter

q) Bucket

r) Mixing box

s) Wheelbarrow

Figure 69: Common masonry tools

Some other tools are less common but extremely helpful to facilitate the job and achieve a good quality of work (Figure 70). Some of these tools are not expensive thus absolutely worth having. For example, a 10 to 12 m transparent water hose which is a cheap but most useful tool to trace a reference level, a screed to level out concrete and to trace levels over a short distance, steel cones for slump tests, chalk powder or a chalk line to mark alignments on slabs, and a big brush to wet bricks or blocks before use.

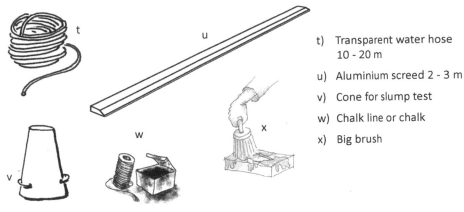

t) Transparent water hose
    10 - 20 m

u) Aluminium screed 2 - 3 m

v) Cone for slump test

w) Chalk line or chalk

x) Big brush

Figure 70: Additional and inexpensive tools

Finally, there are some pieces of machinery that are highly recommended for any small-scale contractor (Figure 71). They require a certain amount of investment but they're very useful. Not only do they facilitate work but they do also increase the quality of the end product. They are:

a) Needle vibrators (vibrating probes) to compact concrete. There should be two diameters of needles available, a thin one for narrow vertical ties or columns, and a thicker one where there is more space. Some training will be necessary to use the needle vibrators correctly. Vibrating has to start from the bottom of the last batch of concrete poured into the formwork, slowly pulling the needle upwards. Care must be taken to limit vibrating to about 15 seconds, as excessive vibrating will cause concrete to separate into cement water and aggregates.

b) A concrete mixer allows for a well mixed concrete with regular characteristics between the various batches (see the previous section for recommendations on how to make good concrete with a mixer).

c) A grinder to cut concrete blocks. Concrete blocks can be difficult to break into the right length with a hammer. In addition, hammering will cause micro-cracks in the blocks which will weaken them. Cutting blocks with a grinder allows for much more precision and will not affect the quality of the block.

d) Vibrating concrete block press. Concrete blocks made without a press and without vibration don't have the correct density and are too weak. Blocks should always be made with a press, then cured properly (see earlier section on the production of concrete blocks).

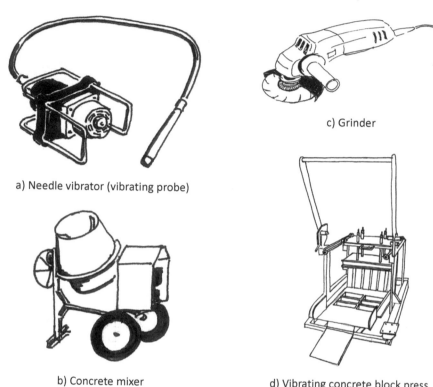

a) Needle vibrator (vibrating probe)

c) Grinder

b) Concrete mixer

d) Vibrating concrete block press

Figure 71: Highly recommended pieces of machinery

Part II:
Confined masonry step by step

Notice:

The measures in the following section are given in metric and in imperial units (in brackets).

Metric units: 1 m = 100 cm = 1000 mm

In order to keep things simple, 1 inch is made to correspond to 2.5 cm (25 mm) and 1 foot to 30 cm.

Sometimes the conversion between the two units does not seem to be correct. This is due to the rounding mentioned above. Multiplications within one unit however are correct.

Example: 15 cm ≈ 6". When multiplied by 16:
16 x 15 cm = 240 cm; 16 x 6" = 96", even if 96" are equal to 244 cm.

There are only a few countries using imperial units where Ø 8 mm rebars are available. For this reason their next higher "equivalent" in imperial units is given (i.e. 3/8 in. which corresponds to 10 mm).

http://dx.doi.org/10.3362/9781780449906.003

CHAPTER 3   SITE PREPARATION

3.1   Clearing the ground

Prepare the construction site by removing shrubs, grass, loose material and debris from former buildings. This has to be done on a more generous portion of the ground than what is actually needed for the house to allow for enough room to work and to install your equipment. Cut trees only if their roots might disturb the foundations.

Trace the house outline plus 2 - 3 m (7 - 10 ft.) all around with chalk or lime.

Remove the top soil until the underneath firm soil is exposed. Set the top soil in a pile off the side. Reuse it around the site for gardening once the construction is complete (Figure 72).

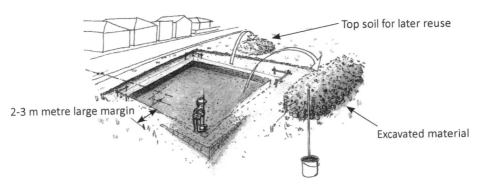

Top soil for later reuse

2-3 m metre large margin

Excavated material

Figure 72: Removal of top soil and excavation

3.2   Levelling the ground

Level the firm soil by using an optical instrument or a long transparent water hose. Fill the hose with water. The water in the hose should always be level. The excavation should be perfectly horizontal (Figure 73).

If there is a sewage network, make sure that the bottom of your excavation is above the level of the sewage collector pipes.

Figure 73: Levelling the ground with a transparent water hose

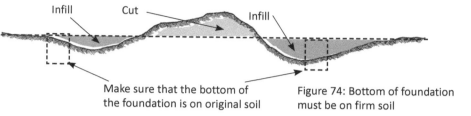

Infill Cut Infill

Make sure that the bottom of the foundation is on original soil

Figure 74: Bottom of foundation must be on firm soil

If there are areas below the desired level, fill them in with excavated soil and compact with a rammer. Don't compact layers of more than 20 cm (8 in.) at a time. Foundations must always be placed on original soil (Figure 74).

3.3 Placing the batter boards

Consult your drawings for the foundation layout to get the maximum width and length of the building. Place stakes at the approximate position of the corners and verify the diagonals to make sure the outline is a rectangle (Figure 75).

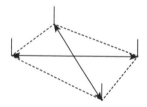

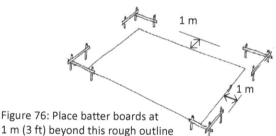

1 m

1 m

Figure 75: Check diagonals to verify right angle

Figure 76: Place batter boards at 1 m (3 ft) beyond this rough outline

Make the boards long enough (approx. 2 m / 7 ft.) so that they can receive the tracing wires for the foundations. Batter boards can be made in one piece and placed diagonally to serve two trenches, or they can be in two pieces, each one placed exactly in front of the foundation to dig (Figures 76 and 77).

Batter board in diagonal

L-shaped batter board

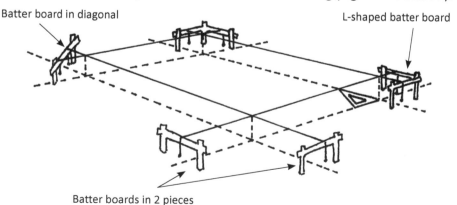

Batter boards in 2 pieces

Figure 77: Placing of batter boards

Place strings (or wires) to outline the trenches to be excavated.

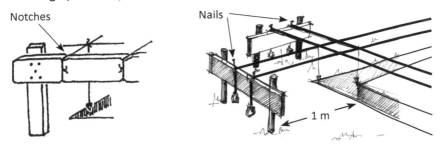

Figure 78 a) and b): Notches or nails to replace strings in the same position

Cut notches or put nails on the batter boards where you have placed the strings (either tie the strings or use weights on the ends of the strings to keep the lines taut). This will allow you to remove the strings when necessary and replace them in the exact same position (Figure 78).

3.4 Tracing a right angle

There are two methods to trace a right angle. Method 1 works well for the placement of the tracing wires, method 2 is more appropriate to trace right angles on a concrete slab.

Method 1: the 3-4-5 rule

A triangle with side proportions of 3 - 4 - 5 defines a right angle. One might use any unit to trace that triangle (Figure 79). So for example, 3 m - 4 m - 5 m or 3 ft - 4 ft - 5 ft lead to the same result. On construction sites where space is often limited, multiples of 30 cm (1 ft.) are most commonly used. (E.g.: 2 x 30 cm = 60 cm, thus 60 - 80 - 100 cm, or 5 x 30 cm = 150 cm, thus 1.5 - 2 - 2.5 m). For more accuracy it is better to use the biggest possible figures.

1. Start by placing the first tracing wire where you want it to be (e.g. alignment of one of the main walls of the house). Place the second wire approximately at a right angle. Mark the point of intersection on both wires with a pen.

2. On the first wire, make a second mark at 90 cm. On the second wire make a mark at 1.2 m.

3. Now let somebody hold a measuring tape with zero against the 90 cm mark. Rotate the second wire until the 1.2 m mark corresponds with the 1.5 m mark on the measuring tape.

3	4	5
30cm	40cm	50cm
60cm	80cm	100cm
90cm	120cm	150cm
1.5m	2m	2.5m
2.1m	2.8m	3.5m
3m	4m	5m

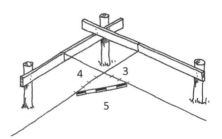

Figure 79: Example: 3 x 30 = 90 cm, 4 x 30 = 120 cm, 5 x 30 = 150 cm

Figure 79 a): Multiplying each basic figure by the same amount maintains the proportions

Repeat the operation in each corner and verify the diagonals. If your corners are all at right angles, the diagonals will have the same length (Figure 80).

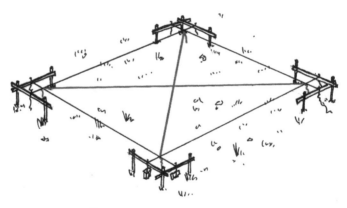

Figure 80: Diagonals of a rectangle have the same lengths

Attention: Same length diagonals do not guarantee that you have right angles, as you can see in the following illustrations (Figure 81). But they are a good technique to verify if you have made your right angles properly.

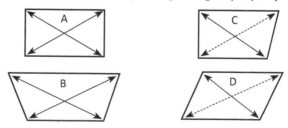

Figure 81: A and B: diagonals of the same length, C and D: diagonals are different.

Method 2: Geometry (to be used on concrete slabs or other hard surfaces)

1. Trace a line (or use the edge of the slab) and draw a semi-circle to mark two points B at the same distance from a centre A;

2. From points B, trace 2 parts of circles with the same diameter in such a way that they cross somewhere (point C);

3. Pass a line through A and C. This line is perpendicular to the baseline (Figure 82).

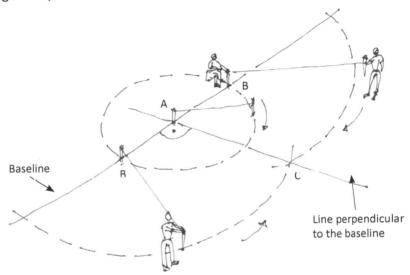

Baseline

Line perpendicular to the baseline

Figure 82: Tracing a right angle with geometry

http://dx.doi.org/10.3362/9781780449906.004

CHAPTER 4   FOUNDATION

4.1   Type of foundation

In the confined masonry system foundations have to be continuous (strip foundation) because they have to carry the load bearing walls (Figure 83). Isolated footings placed under the columns are not correct because confined masonry does not have columns but vertical ties and loadbearing walls.

Figure 83: Foundations must be continuous to support the walls on their whole length

Foundations can be made out of concrete (i.e. normal reinforced or unreinforced concrete or rough concrete with large embedded stones) or out of masonry (stone, bricks or concrete blocks) (Figure 84). If concrete blocks are used, make sure that they are of very good quality so they will not disintegrate in damp ground conditions.

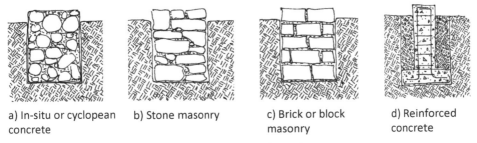

a) In-situ or cyclopean concrete   b) Stone masonry   c) Brick or block masonry   d) Reinforced concrete

Figure 84: Different types of strip footings

Cover the bottom of the trench with a 5 cm (2") layer of lean concrete (cement - aggregate ratio 1:10) to have a clean surface to start with (Figure 85). Where the terrain is soft or of irregular density it might be necessary to add a reinforced concrete base under the footing. This reinforced concrete must be cast onto a layer of lean concrete (Figure 86).

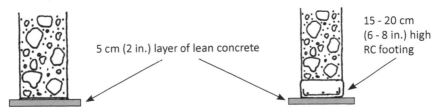

5 cm (2 in.) layer of lean concrete

15 - 20 cm (6 - 8 in.) high RC footing

Figure 85: It is good practice to start with a layer of lean concrete

Figure 86: On irregular soil, foundations are reinforced with a continuous reinforced concrete footing

4.2 Width and depth of foundation

Depth:

Foundations should reach at least 50 cm (20 in.) into firm ground and the whole foundation should have a minimum height of 80 cm (32 in.) (Figure 87). However, minimum depth might be greater because the foundation must go below freezing level (frost would push the foundation upwards).

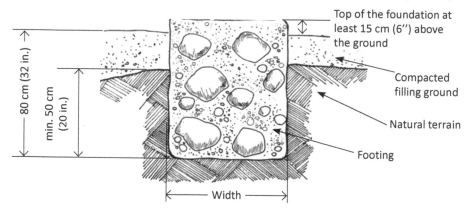

Figure 87: Depth of foundations

Width:

The width of the trenches will depend on the type of soil. Hard soils such as rock or compact gravel are the best foundation soils. Using the soil analysis test, you can determine if the soil is classified as clay or sandy (Figure 88).

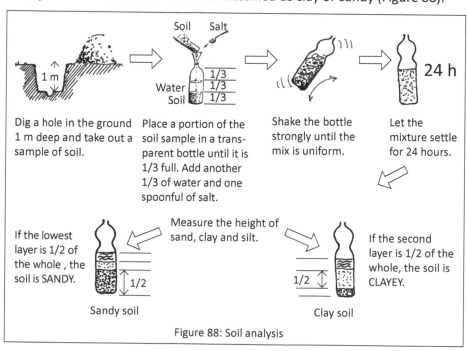

Figure 88: Soil analysis

The footing width for houses up to two stories with bearing walls should be according to Figure 89:

Type of soil	Minimum width of the foundation
Hard soil (rock and compact gravel):	40 cm (16 in.)
Clay soil or clay sand:	50 cm (20 in.)
Sandy soils:	70 cm (28 in.)

Figure 89: Width of foundation according to soil

Also enquire about nearby houses. If they have settled under their weight, then your foundation should be wider and deeper than that of your neighbours.

4.3 Excavating the trenches

Once the width and exact position of the foundation is known the tracing wires can be placed on the batter boards and with the aid of a plumb line the trenches can be traced with chalk (Figure 90). Then the wires are removed to make room to excavate the trenches (Figure 91). Take care when making the vertical walls of the trenches. Trenches must go down to good hard soil and must be free from any organic material such as roots, leaves, garbage, etc. Wet the base of the trench and compact with rammer. The bottom must be level.

Figure 90: Trace trenches with chalk

Deposit the dug out material at least 60 cm from the trenches so it won't fall back into them when it rains.

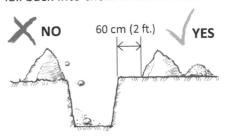

Figure 91: Keep dug out material away from trenches

If you build on a slope, the foundation must be stepped, keeping the bottom of the trench always horizontal (Figure 92).

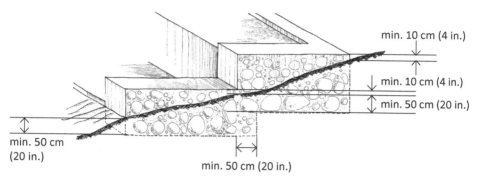

min. 10 cm (4 in.)

min. 10 cm (4 in.)

min. 50 cm (20 in.)

min. 50 cm
(20 in.)

min. 50 cm (20 in.)

Figure 92: Staggered foundation in slopes

4.4 Building the foundations

Placing the vertical reinforcements:

For stone foundations, place the reinforcement bars of the vertical ties on a 5 cm (2 in.) layer of lean concrete at the bottom of the foundation and add spacers between the concrete base and the rebars. Leave a void of 5 cm (2 in.) around the reinforcement cage to be able to pour in concrete once the foundation is complete (Figures 93 & 94).

For brick or block masonry foundations where it might be difficult to leave a void around the vertical rebar cage due to the narrow width of the foundation, one might start the vertical reinforcement at plinth level and connect it well with the plinth reinforcement.

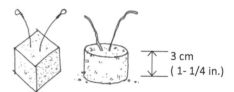

3 cm
(1- 1/4 in.)

Figure 93: Spacers below rebars ensure proper embedding in concrete

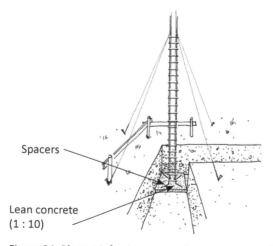

Spacers

Lean concrete
(1 : 10)

Figure 94: Place reinforcing cage on lean concrete and spacers

If foundation work has to be interrupted, leave the ends staggered and rough so that the next bit of foundation can hook properly to the first one. Drench fresh foundations every day and cover them against evaporation (Figure 95).

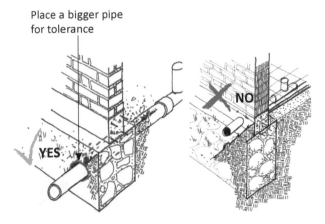

Place a bigger pipe for tolerance

Figure 95: Interrupted foundation work must have staggered and rough ends. Keep fresh foundations moist.

Figure 96: Sewage pipes must pass through the foundation, not through the plinth beam

Sewage pipes must pass through the foundation, not the plinth beams (Figure 96). Leave a bigger hole than needed for the pipe so that there is some tolerance if there is any settlement of the foundation or infill. Don't use empty cement bags to form the void, but use a bigger diameter pipe or make a proper form with boards. Cement bags always end up being crushed and the resulting hole will be too small.

http://dx.doi.org/10.3362/9781780449906.005

CHAPTER 5   VERTICAL AND HORIZONTAL TIE REINFORCEMENTS

Vertical and horizontal RC (reinforced concrete) ties are the essence of confined masonry. They confine the wall elements, making them more resistant to earthquake forces (Figure 97).

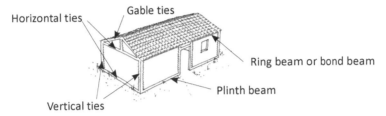

Figure 97: Position and naming of horizontal and vertical ties

5.1 Vertical ties ("tie columns")

The armature of the vertical ties starts at the very bottom of the foundation.

The minimum number of lengthwise rebars is four (vertical ties with only 3 rebars are not permissable).

Spacing of the stirrups:

Stirrups are spaced every 20 cm except for the first and last sixth of the height (H/6) where they are doubled, resulting in a spacing of 10 cm. The first stirrups above the plinth and the last stirrups below the bond beam should be spaced by 5 cm. (Figure 98).

Note: The ties that encase the vertical reinforcement are referred to here as stirrups in order to differentiate them from the concrete elements that are referred to here as ties. These stirrups are often referred to in other publications as ties.

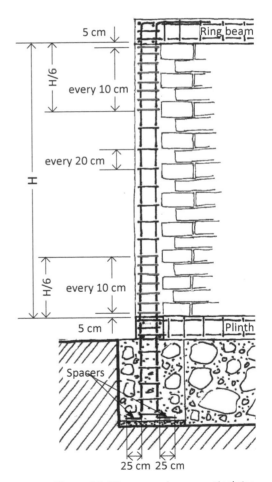

Figure 98: Stirrup spacing on vertical ties

Diameters of rebars:

Minimum diameter for lengthwise (i.e. vertical) reinforcement bars is 10 mm (3/8"). Rebars have to be deformed. If deformed steel cannot be found, the minimum diameter must be increased to 12 mm (1/2").

Stirrup diameter is 6 mm (1/4"). Rebars can be smooth (Figure 99).

Placing of stirrups:

The opening of the stirrups should be alternated along the armature (Figure 99).

Form of stirrups:

Stirrups bent at a 90° angle can get loose during a quake (Figure 100).

Figure 100: 90° stirrups can loosen during a quake. They are not permissable.

Figure 99: Diameter of rebars and placing of stirrups

There are two ways of making solid stirrups which will not open under stress:

- Type 1: Stirrup ends are bent inwards at a 45° angle (often also called 135° angle). The minimum length of the hooks is 6 cm (i.e. 10 x diameter).
- Type 2: Stirrups are bent 1¾ times. Such stirrups are particularly useful in very narrow tie columns (Figures 101 & 102).

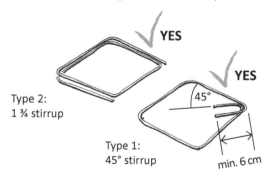

Figure 101: Allowable stirrups: type 1 and type 2

The minimum cross section of tie columns (vertical ties) is 15 x 15 cm but only if stirrups bent at 1 ¾ (type 2) are used.

With normal stirrups (type 1 with 45° hooks) the minimum cross section is 20 x 20 cm (Figure 102).

Foundations:

1. Keep excavated material further away from trenches.
2. A clean trench bottom made with lean concrete.
3. The vertical reinforcement should start below the foundation.

Foundations:

4. Keep a void around the vertical reinforcement at foundation level, so it can be filled with concrete.
5. Foundations for confined masonry must be continuous. Don't forget to place the door reinforcement in the plinth.
6. On soft soil it is preferable to place an RC footing under the foundation.

YES

7

NO

8

NO

9

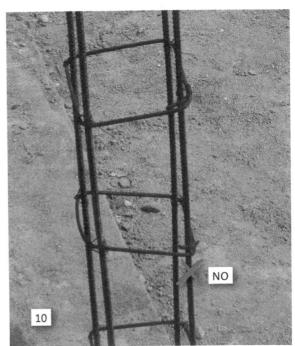

NO

10

Reinforcement and connections:

7. Correct reinforcement of vertical tie, with 45° stirrups and 10 cm (2 in.)spacing near plinth.
8. Stirrups are bent at 90° angles instead of 45° and spacing between stirrups has not been reduced next to the plinth.
9. Don't use salvaged and straightened rebars.
10. Stirrups are too far apart, with no 45° hooks (and one stirrup is even incomplete).

YES

11

YES

12

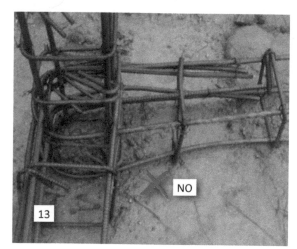

NO

13

Reinforcement and connections:

11. Correct corner connection: rebars from inside to outside.
12. Connection at ring beam level made with L-shaped rebars. Notice the closer spacing of the sirrups next to the vertical tie.
13. With poorly bent rebar it becomes impossible to make straight formwork for the plinth.
14. Inner rebar must never go around the inner corner.

NO

14

15

16

17

19

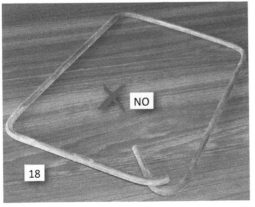

18

Reinforcement and connections:

15. Little hooks are not enough to ensure a solid connection
16. Stirrup spacing of the vertical tie is too large.
17. Insufficient vertical reinforcement and very weak connection to beam.
18. Steel test: if you can't bend a 45° hook without breaking the rebar, the steel is too stiff and must not be used.
19. Little hooks are not sufficient to ensure a solid corner connection.

Spacers:

20. Spacers are necessary to keep reinforcement at the right distance from the formwork.
21. Spacers can be made with 3 cm plastic tubes cut into 3 cm pieces.
22. Or they can be made in a 3 cm high form and cut in squares like a cake.

Masonry:

23. Joints must not be larger than a little finger (12-15 mm / 1/2 inch)
24. Moisten blocks and bricks before use, so they won't absorb the water of the mortar.
25. Fill the vertical joints between the masonry units.
26. Correct size of toothing with bricks.

27

28

30

29

31

Masonry:

27. Correct size of toothing with blocks.
28. With correct size of toothing concrete will penetrate properly.
29. Toothing is too deep. Concrete will not be able to penetrate properly.
30. Concrete has not been compacted / vibrated properly. Honeycombs do not protect the rebars properly and weaken the concrete element.
31. Never place drainage pipes in the vertical ties.

Why are there two types of stirrups?
There are two factors that define the minimum dimension of 45° stirrups:

- The 45° inward bent hooks have to be at least 6 cm long and the position of the hooks should alternate along the armature.
- All rebars, including stirrups, should be covered by 30 mm of concrete to protect them against rust.

When a stirrup is drawn according to these two conditions, and where the inward bent stirrup hooks do not cause an excessive obstacle to the flow of concrete and to its compaction, the minimum dimension for a tie-column becomes 20 x 20 cm.

Thinner tie-columns, say 15 x 15 cm, require smaller stirrups where the central gap can be too narrow for the passage of a needle vibrator. In this case 1 ¾ stirrups are recommended.

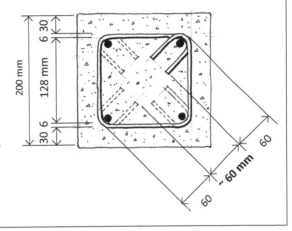

Figure 102: Calculation of a minimum 45° stirrup

The use of 20 x 20 cm vertical ties is recommended, even when 15 cm wide concrete blocks are used. The necessary formwork is not particularly difficult in either case. 15 cm wide vertical ties with 45° stirrups are only possible when the tie is longer than min. 20 cm (8 in.) (Figure 103).

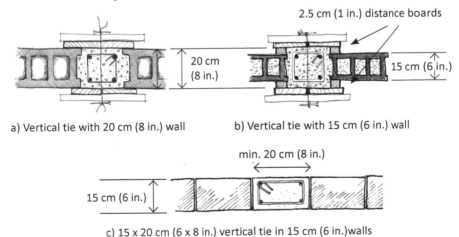

a) Vertical tie with 20 cm (8 in.) wall

b) Vertical tie with 15 cm (6 in.) wall

c) 15 x 20 cm (6 x 8 in.) vertical tie in 15 cm (6 in.)walls

Figure 103: Recommended minimum vertical tie dimensions with 45° stirrups

In countries such as Indonesia where 11 cm (4 ½ in.) wide clay bricks are used for wall masonry, tie columns should not be smaller than 15 x 20 cm (6 x 8 in.) if stirrups with 45° hooks are used. The formwork can be made by adding 2 cm thick vertical distance boards (Figure 104).

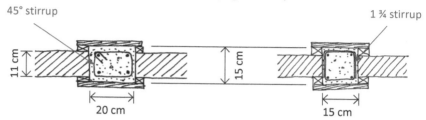

Figure 104: Minimum tie-column dimensions for 11 cm wide walls depend on type of stirrup used

5.2 Horizontal ties (plinth and ring beams)

The plinth beam is the horizontal reinforced concrete (RC) tie placed between the foundation and the wall. It is one of the four basic confining elements that hold a wall together. The ring or bond beam is the horizontal RC tie placed on top of a wall. They both must be well connected with the vertical ties.

For the diameters of the rebars and their spacing the same rules apply as with vertical ties. Note: the number of stirrups near to the vertical ties should be doubled, resulting in a spacing of 10 cm (4 in.)between each other (Figure 105).

With regard to the form, type 1 stirrups with 45° hooks should be used everywhere.

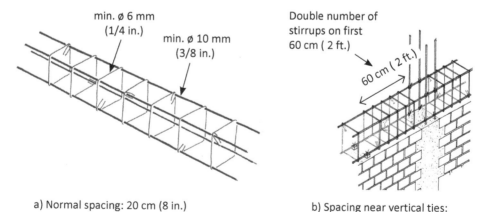

a) Normal spacing: 20 cm (8 in.)

b) Spacing near vertical ties: 10 cm (4 in.)

Figure 105: Diameter of rebars and spacing of stirrups

The width of a plinth or bond beam must be the same as the width of the vertical tie and not less than the wall built on top of the plinth. Minimum width is therefore 15 cm (6 in.) though 20 cm (8 in.) is preferable.

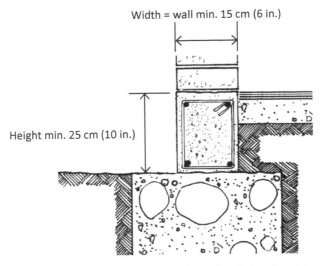

Figure 106: Minimum dimensions of a plinth or bond beam

5.3   Rebar connections: overlapping lengths

Insufficient overlapping lengths of rebars is one of the most common errors made by ironworkers. Correct overlapping lengths are necessary to allow for enough concrete to be able to grip the steel rods and transmit the efforts from one rebar to another. It is like holding two rebars with one, two or four hands: of course, four hands have the strongest grip (Figure 107).

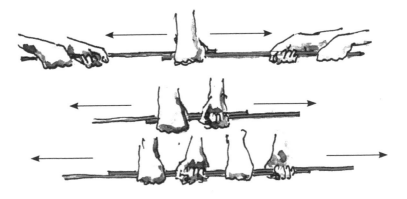

Figure 107: Longer overlapping lengths offer a stronger grip

The minimum overlapping length of straight rebars is 50 times their diameter (Figure 108):

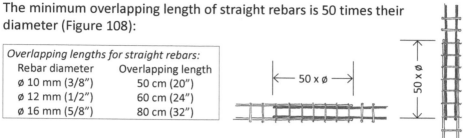

Overlapping lengths for straight rebars:	
Rebar diameter	Overlapping length
ø 10 mm (3/8")	50 cm (20")
ø 12 mm (1/2")	60 cm (24")
ø 16 mm (5/8")	80 cm (32")

Figure 108: Overlapping length for straight rebars = 50 x ø

5.4   Rebar connections at plinth and ring beam corners

To be effective, ties have to be continuous all around the building. Proper tie connections around the corners are extremely important and a frequent source of errors (Figure 109).

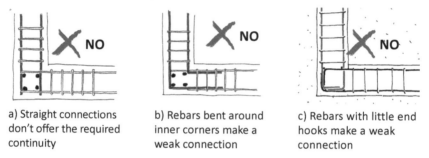

a) Straight connections don't offer the required continuity

b) Rebars bent around inner corners make a weak connection

c) Rebars with little end hooks make a weak connection

Figure 109: Typical and frequent erroneous corner connections (View from above)

Corner connections must follow the *inside to outside rule*. This means that lengthwise rebars on the inside of a plinth have to be anchored on the opposite (external) side of the perpendicular plinth (or bond beam) (Figure 110). The thickness of the concrete (together with the stirrups) will ensure the anchoring.

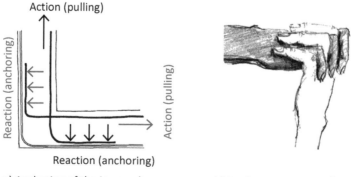

a) Anchoring of the inner rebars

b) A rule easy to remember

Figure 110: The *inside to outside rule* for anchoring the rebars

This result can be achieved either by using hooked, L-shaped or U-shaped connection rebars. L-shaped connectors are easier to use because they require less precision than U-shaped ones (Figure 111).

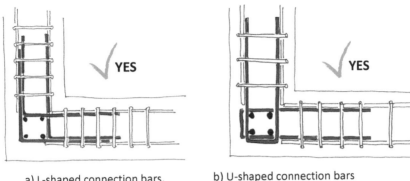

a) L-shaped connection bars,
the easiest way

b) U-shaped connection bars
require much precision

Figure 111: Correct corner connections (View from above)

L-shaped connections can be made in two ways: either the lengthwise rebars of one plinth are bent at the end to enter into the next rebar, or ends are kept straight and L-shaped connectors are added. This second solution allows for easier work (the main armature of the plinth are all made the same way) but requires more steel (Figure 112).

Note that the overlapping lengths must always be at least 50 times the diameter of the rebars.

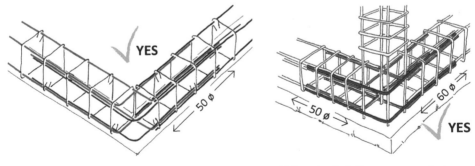

a) Lengthwise rebars of one plinth bent
to enter into the other one

b) Standard plinths are connected
with L-shaped connectors

Figure 112: Two ways to connect plinths with L-shaped rebars

The plinth connections don't have to be at the end of the plinth but can also be made at the middle. This solution allows for a lot of flexibility with regard to the exact length of the plinth reinforcement (which often cause difficulties to the ironworkers) (Figure 113).

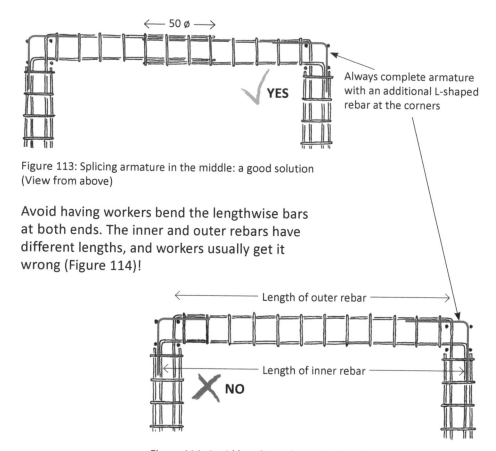

Figure 113: Splicing armature in the middle: a good solution (View from above)

Avoid having workers bend the lengthwise bars at both ends. The inner and outer rebars have different lengths, and workers usually get it wrong (Figure 114)!

Figure 114: Avoid bending rebar at both ends: errors are frequent (view from above)

5.5 T-junctions of plinth and ring beams

The *inside to outside rule* described above also applies to T-junctions. Never bend the rebars around the inner corners and never make connections with straight rebars (Figure 115).

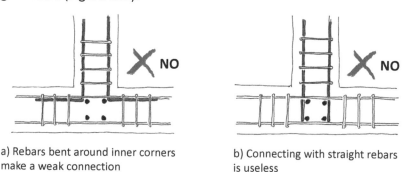

a) Rebars bent around inner corners make a weak connection

b) Connecting with straight rebars is useless

Figure 115: Unacceptable T-connections (View from above)

In order to connect an intermediate plinth beam with a lengthwise plinth beam, bend the end of the rebars at right angles and place them against the outer rebars of the lengthwise plinth (Figure 116).

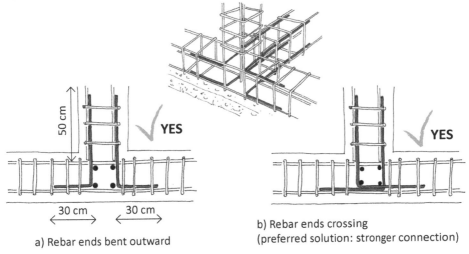

a) Rebar ends bent outward

b) Rebar ends crossing
(preferred solution: stronger connection)

Figure 116: Connecting rebars must be placed against outer rebars. Methods a) and b) are acceptable

5.6 Placing the formwork

The formwork must be solid and well braced.

Do not use wires but small boards to keep the distance of the lateral boards (Figure 117).

Always water the formwork before pouring concrete. Wetting the boards will make them swell and become water-tight. Additionally, it prevents the formwork from absorbing water from concrete.

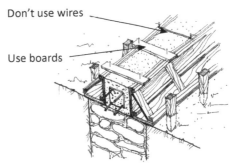

Figure 117: Formwork must be solid

5.7 Preparing spacers

Spacers help to protect rebars against rust by ensuring a sufficient concrete cover. For this reason spacers must have a height of 3 cm (1-1/4 in.) (Figures 118 & 119).

If the stirrups have not been bent with enough precision and there is not enough room for 3 cm spacers on both sides of the rebar cage, use the 3 cm spacers on the outside where the concrete will be exposed to weather.

Rather than not putting spacers on the other (inner) side, you might have to produce an additional batch of 2 cm (3/4 in.) high spacers. Not using spacers is not an option.

Place spacers between the rebar cage and the formwork approximately every 60 to 90 cm (2 - 3 ft.).

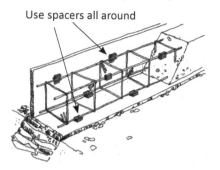

Use spacers all around

Figure 118: Spacers are compulsory

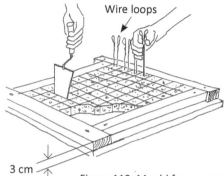

Wire loops

3 cm

Figure 119: Mould for spacers

In many countries spacers can be bought directly from building material suppliers. Make sure that you buy 3 cm (1- 1/4 in.) thick spacers. If they are not readily available, it is quite easy to produce them on a building site.

- Make a 3 cm high mould and fill with rich mortar. Cut the fresh mortar into small 3 x 3 cm pieces and stick a wire loop into each piece. Once hard, break the pieces apart with a hammer.

- Alternatively one may also cut 3 - 4 cm plastic pipes into 3 cm pieces. Fill these with fresh mortar and insert a wire loop. When the concrete has hardened, cut the plastic tubes to remove the spacers (Figure 120).

3 - 4 cm

3 cm
1 - 1/4 in.)

3 cm

3 cm   3 cm

a) Plastic pipe cut into 3 cm bits.   b) Cube shaped spacers 1 - 1/4 " all sides

Figure 120: Two types of 3 cm high spacers

5.8 Pouring the concrete (for the preparation of concrete,
see chapter 2.11)

Pour the concrete and compact it with a needle vibrator (ideally) or poke a rod into the concrete (about 20 times in the same area). In addition knock a hammer against the formwork to help compact the concrete (Figure 121).

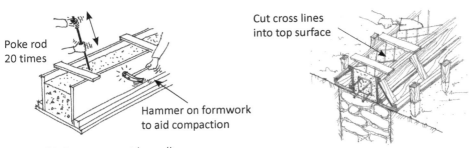

Poke rod
20 times

Hammer on formwork
to aid compaction

Cut cross lines
into top surface

Figure 121: Concrete must be well
compacted

Figure 122: Roughen up top surface of plinth

Finish the upper surface of the concrete by roughening it up with cross lines made with a trowel. This will facilitate the adherence of the mortar. The top surface of the plinth must be perfectly horizontal but not smooth (Figure 122).

Finally, to prevent the freshly cast concrete from drying out, cover it with a tarpaulin or empty cement bags. Water the concrete twice a day for the first week and cover it again each time.

http://dx.doi.org/10.3362/9781780449906.006

CHAPTER 6   WALLS

In confined masonry, walls are the most important elements. They ensure the earthquake resistance of a building. The walls are confined by horizontal and vertical ties which hold them together under earthquake stress. It is vital that the masonry work of the walls is done with quality materials and with utmost care.

Masonry units (concrete blocks, bricks) must be strong and regular in shape. See section 2.10 for more details on how to select proper masonry units. For recommendations on how to prepare mortar, see section 2.11.

6.1   Width of masonry unit and maximum wall height

The minimum wall thickness for brick or full concrete block masonry is 11 cm (4 ⅜ in.). For hollow concrete block masonry the minimum wall thickness is 15 cm (6 in.) (Figure 123).

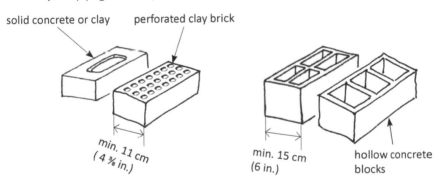

Figure 123: Masonry unit minimal width

The width of the masonry unit defines the maximum height of the walls (Figure 124). It is recommended that the height of a wall does not exceed 22 times its width.

In countries where very good solid bricks or blocks are available, the maximum ratio may go up to 1/25.

However, the maximum height of a wall is limited to 3 m (10 ft.) independent of the width of the masonry.

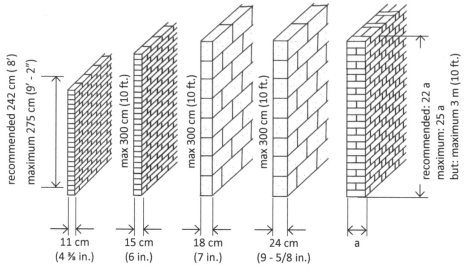

Figure 124: Wall height depends on wall width

6.2 Preparing the masonry units by watering them

To avoid masonry units absorbing water from the mortar, moisten the units (mortar with insufficient water will dry instead of harden properly, leading to masonry work with very weak bonds between the units). Therefore, a few hours before building the walls, submerge the bricks in water for 20 minutes, then take them out and let them sit to get a dry surface (Figure 125).

Concrete blocks: Soak them by dipping them into a bucket full of water.

Alternatively, you may use a brush to get the blocks wet.

Submerse completely

Figure 125: Soak masonry units before use

6.3 Placing the masonry units

Before placing the first layer, plan the layout of the units. Place the bricks or blocks without mortar to determine the placement pattern. To avoid vertical "cracks", units should overlap by a third to a half of their length. Always place the masonry units one full course (row) at a time. Don't stack them diagonally as it will be difficult to maintain horizontal continuity and vertical plumbness (Figure 126).

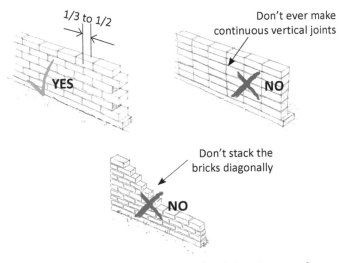

Figures 126: Masonry units must be staggered and placed course after course

Double layer brick walls may be built with either the English or the Flemish bond pattern (Figure 127).

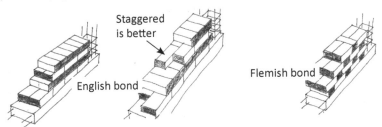

Figures 127: Bond patterns for brick walls

6.4 Toothing or dowels

Wall ends towards the vertical ties have to be made in a zigzag pattern called toothing. These dents (between 5 and 12 cm / 2" to 5") allow for a good connection between the wall and the confining concrete element.

For concrete block walls, break off one or two cells of the last blocks to get 2/3 or 1/3 blocks (Figure 128).

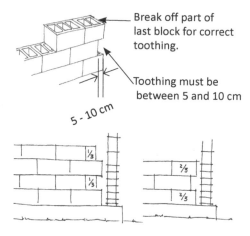

Break off part of last block for correct toothing.

Toothing must be between 5 and 10 cm

5 - 10 cm

Figures 128: Using 1/3 and 2/3 concrete blocks

Leaving 3.5 cm (1 ½ in.) between the last block or brick and the vertical reinforcement allows for concrete to cover the rebars correctly, protecting them against rust. 3.5 cm is about the length of the last joint of your thumb (Figure 129).

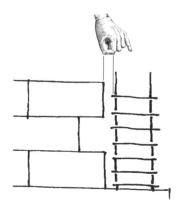

While the toothing method is preferable, there are countries where workers are more used to using dowels. However, there is a contradiction between the requirement that mortar beds should not exceed 15 mm in thickness, and the fact that dowels should be covered with enough mortar to protect them properly against rusting. We therefore recommend that tests are made to see if dowels can be placed properly.

Figure 129: Distance brick/block to rebar is 3.5 cm (1 1/2 in.)

Dowels should be made with 6 mm (1/4") rebars and placed in pairs every 2 layers of blocks or 4 layers of bricks. The dowels should reach 50 cm (1' - 8") into the walls (Figure 130).

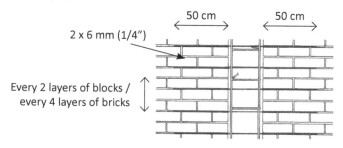

50 cm   50 cm

2 x 6 mm (1/4")

Every 2 layers of blocks / every 4 layers of bricks

6.5   Laying the bricks

Place vertical timber boards. These boards will ensure the verticality of the wall and allow for the fixing of the strings which will help with the alignment and level of each layer of bricks.

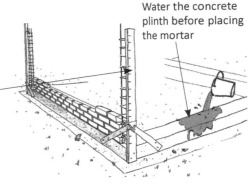

Water the concrete plinth before placing the mortar

Place mortar uniformly over the plinth using a bricklayer's trowel. Bricks should be saturated but surface dry. Place the bricks over the mix and verify that their edges touch the strings that connect the guides (Figure 130).

Figure 130: Vertical boards for level strings

A mortar bed should not be allowed to sit on the plinth beam or bricks for longer than 15 minutes. Once the mortar bed is placed, the bricks should be quickly placed to ensure a good bond (Figure 131).

Before placing the next layer, fill the vertical joints completely.

To better set the brick, hit it lightly with the handle of the trowel.

Figures 131: Placing the mortar and bricks

Do not make joints thicker than 10 - 15 mm (1/2 in.) (the width of a little finger) (Figure 132). Joints that are too thick will weaken the wall.

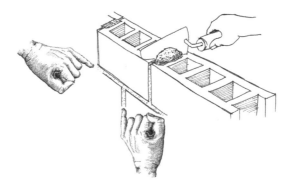

Figure 132: Maximum joint thickness is 10 to 15 mm (1/2 in.)

6.6   Completing a day's work

When the mortar joint is hard enough to make a thumbprint, rake the horizontal and vertical joints by compressing the mortar with a rounded tool (e.g. a bit of electrical pipe).

Do not raise the wall more than 1.2 m (4 ft.) high each working day (Figure 133). If you raise the wall higher in one day, it may go out of plumb because the mortar mix is still fresh.

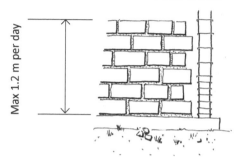

Figure 133: Maximum 1.2 m per day

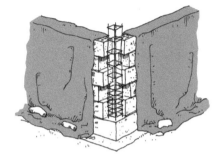

Figure 134: Keep the new walls moist

Once the walls are up, do not forget to cure them by pouring water over them 3 times a day for 7 days and/or by covering them with a tarpaulin (Figure 134).

6.7   Integrating technical installations

Never break into the finished wall to place pipes. Hammering into a wall with a chisel will create micro-cracks in the adjacent blocks and weaken the whole wall (Figure 135).

Figure 135: Never break up the walls for conduits

Electrical pipes

The best way to place electrical conduits without weakening the wall is to leave them visible on the surface (Figure 136).They can also be mounted on the rough wall and then covered with plaster.

This however has the disadvantage that the pipes are hidden and there is a risk of hammering a nail into a live wire.

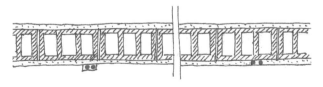

Figure 136: Left: visible conduit; right: conduit integrated in the plaster

If the conduits have to be
hidden in the wall you may leave a void
around the pipes while raising the wall
(Figure 137). These voids will later be filled
with rich mortar or concrete.

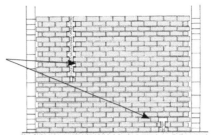

Figure 137: Leaving voids around the
pipes while raising the wall

Finally, for hollow concrete block masonry there
is another option: the hollow blocks may be laid
over the pipes while building the wall (Figure 138).
This is certainly the most elegant, but also the most
complicated solution as the free conduits may be
leaning all over the place before they can be
integrated into the block wall.

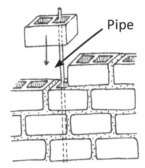

Figure 138: Pipes integrated
into hollow concrete blocks

Water and sewage pipes

The recommended way to place water and sewage pipes is to create a service
shaft (Figure 139). Avoid placing the pipes in the walls. Pipes in walls can ruin
the bracing effect of the wall.

Place bigger pipes in
service ducts (service shafts)

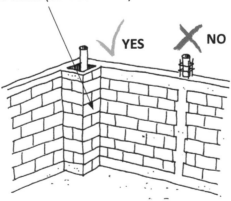

Figure 139: Creating service shafts
for bigger pipes

If placing the pipes into the wall cannot be avoided, put them in the right place before you build the wall (Figure 140). Leave the wall ends toothed and place #8 wires or 6 mm (1/4 in.) bars every 3 layers on both sides.

Also wrap the pipes with #16 wires (2 mm).

Place the formwork bit by bit, pour the concrete and compact well.

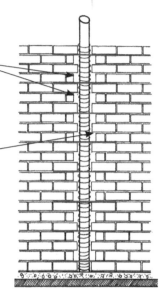

Figure 140: Correct integration of bigger pipes into walls

http://dx.doi.org/10.3362/9781780449906.007

CHAPTER 7   COMPLETING THE VERTICAL TIES

With the foundation, plinth beam, and walls (up to 1.2 m) built, the vertical concrete ties are ready to be poured. Pouring the tie-columns in incremental segments of not more than 1.2 m (4 ft.) at a time allows for good compacting of the concrete, particularly when no needle vibrator is available.

7.1   Placing the formwork

The formwork of the tie-columns should be watertight to avoid leaking of the concrete and the creation of gravel nests (honeycombs). To achieve this, use good timber boards with straight edges. Place the boards against the masonry or, if the wall is thinner than 15 cm (6 in.), add distance boards to get a vertical tie of at least 15 cm (Figure 141). Hold the formwork together with through-wires and brace. it well.

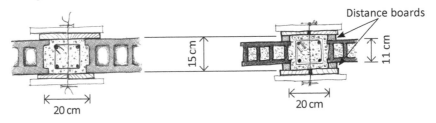

a) Vertical tie with ≥15 cm wide wall   b) Vertical tie with <15 cm wide wall

Figure 141: Minimum vertical tie-column section is 15 x 20 cm

Don't make the formwork for the vertical ties higher than 1.2 m (4 ft.) at a time (Figure 142). This will allow you to clean it properly before pouring the concrete. Also, short formwork will make it much easier to compact the concrete (particularly when using a steel rod rather than a needle vibrator) and it lets the concrete penetrate correctly into the toothing.

Make sure to place enough spacers all around the reinforcement cage. Add some spacers above the pouring joint to keep the rebar in the right place for the next bit of pouring.

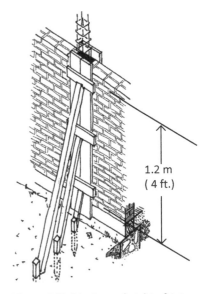

Figure 142: Maximum height of 1.2 m allows for good compacting

83

If you use seismic bands (see section 9.2), place that horizontal formwork too and pour the concrete for both the vertical tie and the seismic band at the same time (Figure 143).

Clean the bottom of the formwork and water the boards thoroughly before pouring the concrete. This will allow the boards to swell and make the form watertight.

Figure 143: Pour concrete of 'tie column' and seismic band at the same time

7.2   Pouring the concrete

Pour the concrete in layers up to 60 cm (2 ft.) at a time and compact. Compacting is important as it makes the air trapped in the concrete escape, leaving a denser (and better) concrete.

If you use a needle vibrator, get it down quickly to the bottom of the layer and pull it up slowly. Don't leave the needle in the same place for more than 5 to 10 seconds. Longer vibrating may disintegrate the concrete by separating gravel and cement. Do not touch the formwork or the steel reinforcement with the needle as it might damage them.

If you compact by hand, poke a steel rod (a thick rebar) 10 to 15 times down to the bottom of the last layer of concrete (Figure 144). At the same time, let a colleague hammer against the formwork. Remember though that compacting without a needle vibrator is less effective.

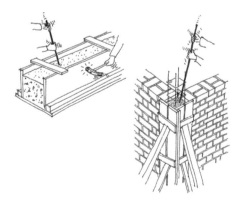

Figure 144: Manual compaction: Poke a steel rod 15 to 20 times into the concrete layer and hammer against the formwork

Figure 145: Water concrete 3 times a day for 7 days

Use concrete within 90 minutes from the time of mixing. Don't refresh concrete which has started to harden. And NEVER add water to make the concrete flow more easily. Concrete with too much water will never be strong.

7.3   Curing the concrete

Proper curing of concrete is very important. Concrete must not *dry* but *harden*. For this chemical reaction to happen it needs water. Curing means keeping the moisture in the concrete for as long as possible.

Leaving the formwork for several days helps to reduce evaporation.

After removal of the boards, check the concrete. If it has a lot of honey-combing (gravel nests), immediately break and remove the concrete, carefully clean the steel bars, replace the formwork and pour the concrete again.

After the removal of the formwork, cure the concrete elements by watering them 3 times a day for at least 7 days (Figure 145). Cover with a plastic sheet or wet cloth in a hot or dry climate (where temperatures can exceed 30°C).

CHAPTER 8   REINFORCEMENT OF OPENINGS

All but the smallest openings must be reinforced with reinforced concrete bands on all four sides. This reinforcement should be anchored to the tie-beams and tie-columns through reinforced concrete bands. There are two ways to do that.

8.1   Using vertical bands

The window and door reinforcements are tied to the plinth and ring beams through tie-columns or vertical reinforced bands. This method offers the possibility to create more confined shear walls.

While it is more complex in its realization and planning (masons have to interrupt their masonry every time they meet a vertical reinforcement, and planners have to do some good thinking), it allows for the creation of enough shear walls in situations where a sufficient length of shear walls in a facade might otherwise be difficult to achieve.

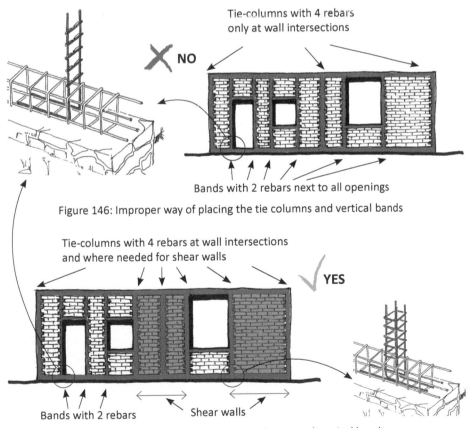

Tie-columns with 4 rebars only at wall intersections

NO

Bands with 2 rebars next to all openings

Figure 146: Improper way of placing the tie columns and vertical bands

Tie-columns with 4 rebars at wall intersections and where needed for shear walls

YES

Bands with 2 rebars   Shear walls

Figure 147: Correct way to place the tie columns and vertical bands

There are 3 rules which have to be observed and balanced against each other (Figures 146 and 147):

1. The total length of all shear walls in each facade has to be at least 50% of the length of the whole façade.
2. Only confined walls with a height to length ratio of 1.5 can be used as shear walls (i.e. if the wall height is 3 m, only confined walls of 2 m or longer can be considered).
3. Each shear wall must be confined with vertical tie columns containing 4 vertical rebars.

Example 1

In the following example the full wall panels (in dark grey) between the main tie-columns are less than half the total length of the facade in the E-W direction, and do not exist in the N-S direction.

In order to increase the total length of shear walls in the facade, some of the remaining walls can be used under the following conditions:

1. Their length to height ratio is bigger than 1.5.
2. They are confined with 4 rebar tie-columns.

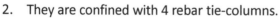

$$L^1 + L^2 > L/2$$

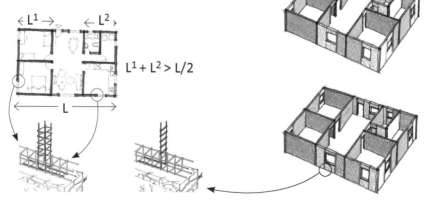

The rest of the vertical bands around the openings can be made with 2 rebars.

Example 2

In example 2 there are no full wall panels between the main tie-columns. Shear walls can only be created by placing 4-rebar tie columns next to the openings (in red).

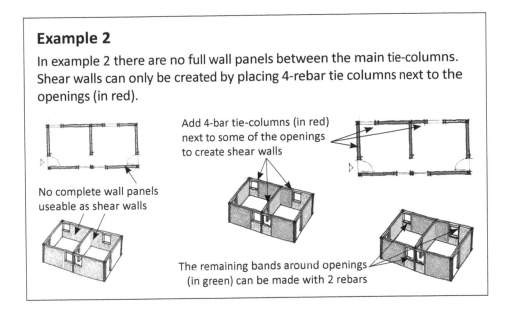

Add 4-bar tie-columns (in red) next to some of the openings to create shear walls

No complete wall panels useable as shear walls

The remaining bands around openings (in green) can be made with 2 rebars

8.2   Using horizontal 'seismic bands'

The reinforcement of the openings is tied into the horizontal bands spanning from tie-column to tie-column (Figure 148). This solution is preferred by masons as it is much easier to implement. In addition, the use of seismic bands enforces the rule of pouring concrete into the tie-columns in well managed segments of a maximum height of 1.2 m (4 ft.) at a time.

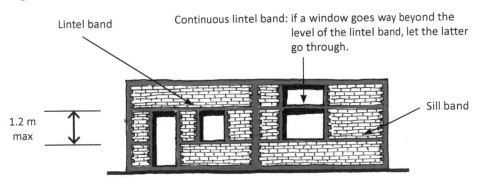

Lintel band

Continuous lintel band: if a window goes way beyond the level of the lintel band, let the latter go through.

Sill band

1.2 m max

Figure 148: Horizontal 'seismic bands' to reinforce openings

Seismic bands are thinner than tie-beams (Figure 149). They only contain two lengthwise rebars. The bands are placed at the sill and lintel level, even in walls without openings. They help to reduce the size of the masonry panels and reinforce the walls, particularly if there is poor workmanship in the masonry or poor quality materials are used.

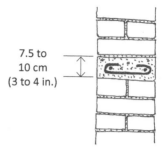

7.5 to 10 cm (3 to 4 in.)

Figure 149: Seismic band

8.3   Connection details of seismic bands

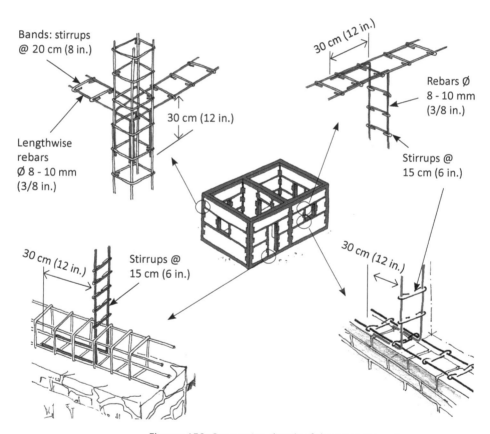

Bands: stirrups @ 20 cm (8 in.)

Lengthwise rebars Ø 8 - 10 mm (3/8 in.)

30 cm (12 in.)

30 cm (12 in.)

Rebars Ø 8 - 10 mm (3/8 in.)

Stirrups @ 15 cm (6 in.)

30 cm (12 in.)

Stirrups @ 15 cm (6 in.)

30 cm (12 in.)

Figures 150: Connection details of the seismic bands

Important: don't forget to place the vertical band for the door reinforcement into the plinth reinforcement before completing the plinth (Figure 150).

8.4 Lintel size

For windows up to 90 cm (3 ft.) the seismic band can be used directly as a lintel, with no need for further reinforcement bars.

Windows between 90 cm (3 ft.) and 150 cm (5 ft.) need additional lintel reinforcement (Figure 151).

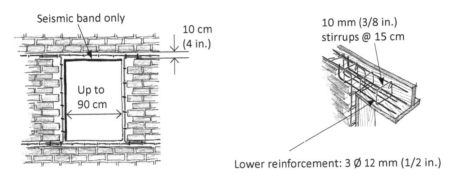

Seismic band only

10 cm (4 in.)

Up to 90 cm

10 mm (3/8 in.) stirrups @ 15 cm

Lower reinforcement: 3 Ø 12 mm (1/2 in.)

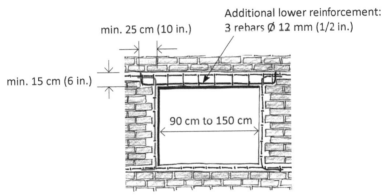

Additional lower reinforcement: 3 rebars Ø 12 mm (1/2 in.)

min. 25 cm (10 in.)

min. 15 cm (6 in.)

90 cm to 150 cm

Figures 151: Lintel height and reinforcement depend on a window's width

In confined masonry construction, don't make windows larger than 1.5 m (5 ft.).

http://dx.doi.org/10.3362/9781780449906.009

CHAPTER 9   RING-BEAMS (TIE-BEAMS, HORIZONTAL TIES)

Horizontal ties are reinforced concrete elements that confine masonry walls at the bottom and at the top. The bottom ties are called 'plinths' or 'plinth beams' and the top ties are the 'tie-beams' or 'ring-beams'. The reinforcement for plinths and ring-beams is basically the same, as are the rules for corner connections and overlapping lengths (Figure 152).

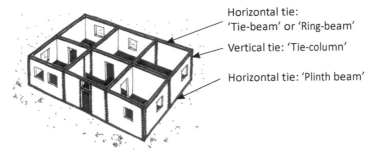

Horizontal tie:
'Tie-beam' or 'Ring-beam'

Vertical tie: 'Tie-column'

Horizontal tie: 'Plinth beam'

Figure 152: Vocabulary for various concrete tie elements

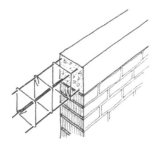

Figure 153: Single ring-beam as found with light-weight roofs

Figure 154: Ring-beam integrated into slab

9.1   Ring beams with light-weight roofs

Ring beams in houses with a light-weight roof (Figures 153 and 154) have a double function: they confine the masonry walls to ensure good in-plane resistance, and they ensure the resistance of the walls against out-of-plane forces (Figure 155). Given the absence of a rigid diaphragm (slab or braced timber floor) they can only resist these forces through their own horizontal rigidity. The larger they are, the stiffer they become and the better they can prevent out-of-plane failure.

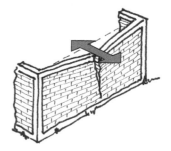

Figure 155: Out-of-plane forces bend the ring-beam and break the wall

In order to ensure a sufficient rigidity of the ring beam, its width (b) must be at least 20 cm (8 in.) or L/20, whatever is larger (Figure 156). L is the distance between adjacent transverse walls.

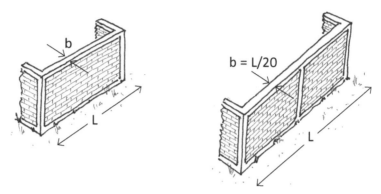

Figure 156: Width of bond-beam = L/20

Minimum reinforcement for all tie-beams is 4 Ø 10 mm (3/8 in.) steel bars with 8 mm (or 3/8 in.) stirrups. (Figures 157 and 158). Stirrup spacing is measured from the reinforcement of the vertical ties. The stirrup spacing from each end is as follows:

- 1st stirrup is placed at 5 cm (2 in.) from end of the beam
- The 5 stirrups after the first are placed at 10 cm (4 in.) intervals
- All other stirrups can be placed at 25 cm (10 in.) intervals.

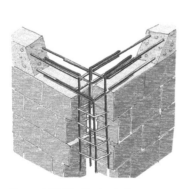

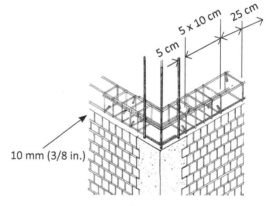

Figure 157: Vertical reinforcement must be bent properly into the bond beams

Figure 158: Reinforcement of tie-beam

http://dx.doi.org/10.3362/9781780449906.010

CHAPTER 10   FREE-SPANNING BEAMS

A free spanning beam does not confine a wall panel and therefore is mostly load-bearing. Its steel reinforcement is similar to that of the tie-beams apart from the maximum distance between stirrups which is 20 cm (8 in.) (Figure 159). Connect well to the ring-beam.

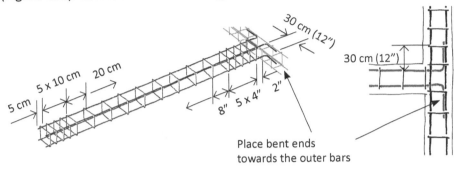

Figure 159: Minimum reinforcement of free-spanning beam

There are two types of free-spanning beams: connecting beams and supporting beams (Figure 160).

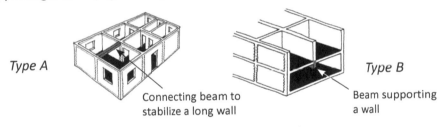

Figures 160: Two types of free-spanning beams

Type A: Connecting beams

This type of beam does not carry any weight, but is placed to hold a long wall in place where no internal wall can accomplish this function.

Type B: Supporting beams

This type of beam is load bearing and integrated in the slab. It reinforces the slab where it carries a partition wall of the upper floor that does not have a corresponding partition wall at the lower floor.

If possible, avoid supporting beams, since walls should always be placed on top of other walls. But if these beams cannot be avoided, there are two ways to make them.

- Deep beams, being lower than the slabs;
- Flat beams, of the same height as the slab.

10.1   Deep beams

Deep beams project under the ceiling (FIgure 161). They are used when the partition wall above is particularly heavy. The minimum depth of the beam is equal to its free span divided by 14.

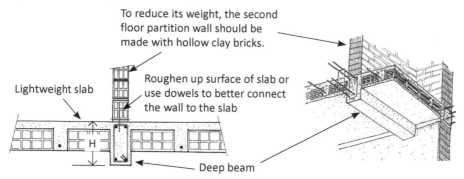

To reduce its weight, the second floor partition wall should be made with hollow clay bricks.

Lightweight slab

Roughen up surface of slab or use dowels to better connect the wall to the slab

Deep beam

Figures 161: Deep beams: H = free span divided by 14

10.2   Flat beams

Flat beams are the same depth as the slab (Figure 162). Flat beams should be designed to have a short free span no greater than 4 m (13 ft.). 8 mm (or 3/8 in.) stirrups can be used for short flat beams. The spacing of the stirrups is the same as all other beams.

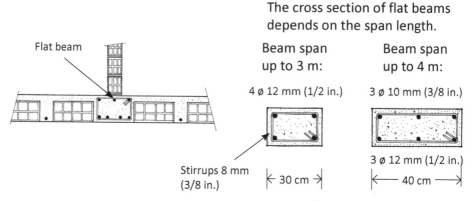

The cross section of flat beams depends on the span length.

Flat beam

Beam span up to 3 m:

Beam span up to 4 m:

4 ø 12 mm (1/2 in.)

3 ø 10 mm (3/8 in.)

3 ø 12 mm (1/2 in.)

Stirrups 8 mm (3/8 in.)

← 30 cm →

← 40 cm →

Figures 162: Dimension and reinforcement of flat beams

Remember: minimum concrete covering for the rebars is 3 cm. In free-spanning beams this is particularly important to ensure fire as well as corrosion protection.

Vertical ties (tie columns):

1. Don't break the walls to insert electrical pipes. Place them on the surface or in the plaster.
2., 3. Vertical ties must not be smaller than 20x20 cm (8x8 in.) . If the walls are thinner, add spacer boards to the formwork (Indonesia).
4. Tie column cast at 20x20 cm with a 10 cm garden wall.

5

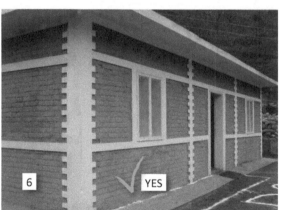

6 YES

Seismic bands next to openings:
5. Two storey building with vertical and horizontal bands (Haiti).
6. Horizontal bands in brick masonry (Pakistan).
7. Horizontal bnds in concrete block masonry (Haiti).

7 YES

YES

8

YES

9

YES

10

Seismic bands around openings:

8. Insert the bent rebar ends of the band reinforcemment into the tie column. The bent part must be 30 cm (1 ft.) long.
9. Band stirrups are placed at 20 cm (8 in.) interval.
10. Cast the bands together with parts of the tie columns.
11. Use the same kind of bands vertically to confine the windows.

YES

11

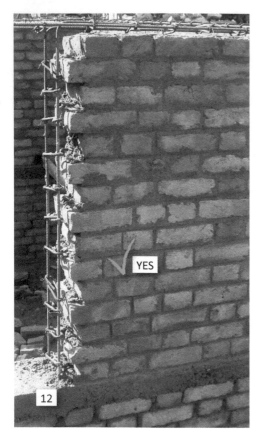

Seismic bands around openings:

12. Use the same kind of bands vertically to confine the windows.
13., 14. Pour the tie columns in segments of 1 to 1.2 m (3'-4" to 4 ft.). This will allow you to compact the concrete more easily.

10.3   Rebar splices in beams

Be careful when you splice reinforcement bars along one beam (Figure 163). Upper reinforcement bars must be cut at the center of the beam span. Lower reinforcement bars must be cut in the first third of the beam.

The overlapping length of the rebars is 60 times their diameter: for 12 mm (1/2 in.) bars it should be 72 cm (30 in.), for 10 mm (3/8 in.) bars it should be 60 cm (2 ft.).

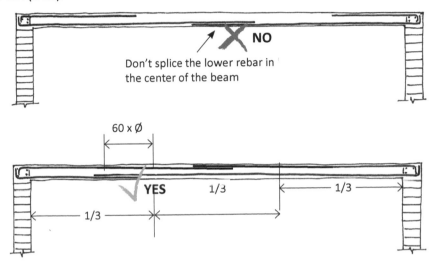

Figures 163: Lower rebars must be spliced in the first third of the span, upper rebars are spliced in the center

http://dx.doi.org/10.3362/9781780449906.011

CHAPTER 11   SLABS

Slabs that are well connected with the ring-beam and the walls are an essential part of the confined masonry technique. They provide rigidity to the whole building. There are two types of slabs in use, depending on a country's traditions: 'full concrete slabs' or 'pan and joist slabs', also known as 'lightweight slabs'.

The following paragraphs provide some basic rules for the construction of simple rectangular slabs of limited dimensions used in simple housing. For anything beyond that, whether with regard to shape or span, contact a structural engineer for appropriate instructions.

Span

The thickness of a slab depends on its span. The span is the distance between two opposing walls on which the slab will be laid. The primary (main) reinforcement of a slab will always be put in the direction of the shorter span. For slabs with a length to width ratio of less than 1.5 / 1, the primary reinforcement will be used for both directions.

The shorter span of the biggest room defines the slab thickness of all rooms. Even if smaller rooms need thinner slabs it is more practical to maintain the same slab thickness for all the rooms (Figures 164, 165, 166).

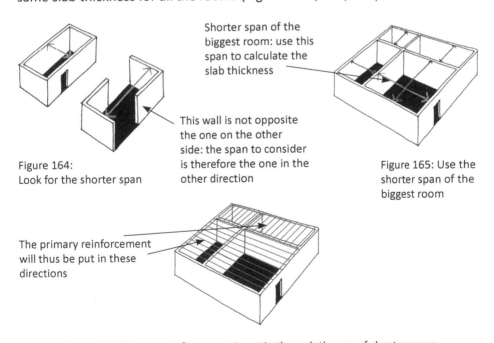

Shorter span of the biggest room: use this span to calculate the slab thickness

This wall is not opposite the one on the other side: the span to consider is therefore the one in the other direction

Figure 164:
Look for the shorter span

Figure 165: Use the shorter span of the biggest room

The primary reinforcement will thus be put in these directions

Figure 166: Primary reinforcement is put in through the use of shorter spans

103

If there is only one room, the slab covering is called a single span slab (Figure 167). Its minimum thickness is span/20. A slab spanning over several rooms is called a continuous slab and its minimum thickness is span/24. In this manual however we recommend using a standard thickness of span/20 for all slabs.

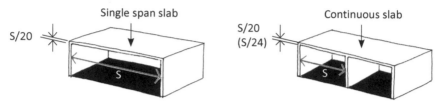

Figures 167: Single span slab and continuous slab

11.1 Formwork

The formwork for slabs has to be sturdy and made with great care because it has to carry a heavy load. A reinforced concrete slab of 18 cm (7 in.) weighs about 450 kg/m² (90 psf). Pan and joist slabs ('lightweight' slabs) are a little lighter, but are still heavy despite their name.

The rules for the construction of the formwork are similar for both types of slabs (Figures 168 and 169):

- Posts must be straight and placed vertically.
- Ground must be levelled and the posts placed on this even surface. No jacking up with bricks or stones is admitted.
- Posts must be braced.
- Horizontal boards must be of good quality and straight-edged to ensure watertightness.

The only difference between the formwork for full and lightweight slabs is that with the latter, horizontal boards are needed only under the joists.

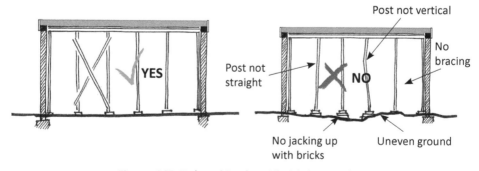

Figures 168: Do's and Don'ts with slab formwork

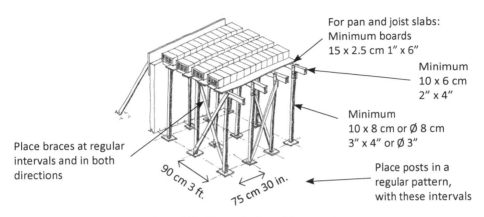

For pan and joist slabs:
Minimum boards
15 x 2.5 cm 1" x 6"

Minimum
10 x 6 cm
2" x 4"

Minimum
10 x 8 cm or Ø 8 cm
3" x 4" or Ø 3"

Place braces at regular intervals and in both directions

Place posts in a regular pattern, with these intervals

90 cm 3 ft.

75 cm 30 in.

Figure 169: Dimensioning of formwork

11.2 Full concrete slabs

Single span slabs

The reinforcement of a single span slab is placed at a distance of 3 cm (1-1/4") from the lower part of the slab. The primary reinforcement goes in the direction of the shorter span and must be well connected with the ring-beam. This means that the ends of the rebars must be bent at a right angle into the ring-beam (Figures 168 and 169).

The secondary reinforcement, usually called shrinkage or temperature reinforcement, is placed in the other direction, on top of the primary rebars.

Don't confuse 'secondary reinforcement' with 'top reinforcement' (negative moment reinforcement) in continuous slabs. Top reinforcement is placed in the upper part of a slab, above walls.

Primary reinforcement Secondary reinforcement / Shrinkage reinforcement / Temperature reinforcement

Straight rebar ends are not admitted

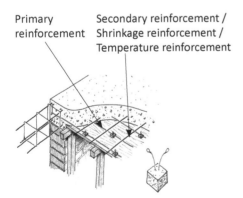

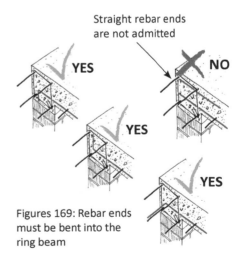

YES

NO

YES

YES

Figure 168: Primary reinforcement is placed first

Figures 169: Rebar ends must be bent into the ring beam

105

With spans of more than 3 m (10 ft.) it is better to bend the rebar ends into a U-shape and let them return 1/6th of the span length (Figure 170).

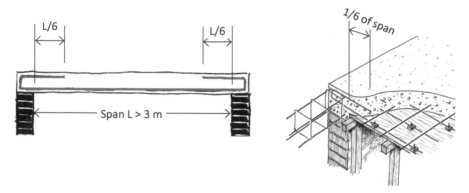

Figures 170: Recommended: U-shaped rebar ends for spans above 3 m

Continuous slabs

The weight of a slab makes it inclined to deform. Because concrete is not good at resisting tension caused by bending, cracks form on the external parts of the curves. To avoid deformation and cracks, rebars have to be placed in these areas (Figure 171).

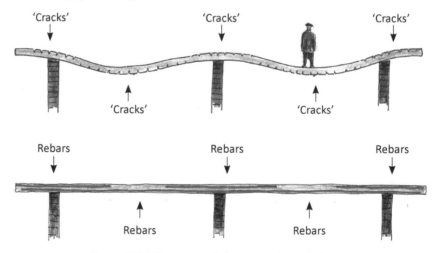

Figure 171: Why upper reinforcement is needed

The lower reinforcement in continuous slabs is the same as in single span slabs. However, where the slab crosses a wall, upper reinforcement is needed in the areas indicated in Figure 172. To keep things simple you may use the same rebar sizes and spacing as with the lower reinforcement.

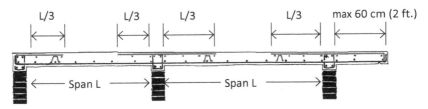

Figure 172: Additional upper reinforcement in continuous slabs

Use 'chairs' to keep the upper reinforcement in place (Figure 173). In over-hanging slabs, create a water drip (Figure 174) to prevent water from running along the lower face of the slab towards the wall. To do that, place a little triangular wood fillet in the formwork, at 2.5 cm (1 in.) from its end.

Figure 173: Place upper reinforcement on chairs

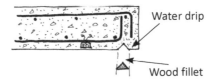

Figure 174: Create a water drip

Reinforcement for full concrete slabs

The quantities, sizes and spacing of the reinforcement bars in a full concrete slab depend on the steel quality which is expressed in psi (grade) or N/mm^2. Grade 60 should be used if available (Figure 175). Grade 40 steel is of lower quality and less expensive. This type of steel is often found in poor countries, together with low quality steel produced from molten scrap steel (e.g. from old cars). If the steel quality is unknown, it is best to use the guidance provid-ed in the table (Figure 176) for grade 40.

Figure 175: Steel grade indications on rebar exist only in certain countries

Rebars with steel grade 40 = 275 N/mm²

In metric units:

Span L	Slab thickness h	Primary reinforcement	Secondary reinforcement
up to 3.0 m	15 cm	Ø 10 mm @ 12 cm	Ø 8 mm @ 15 cm
3.0 m to 3.6 m	18 cm	Ø 12 mm @ 15 cm	Ø 8 mm @ 20 cm
13.6 m to 4.2 m	20 cm	Ø 12 mm @ 15 cm	Ø 8 mm @ 20 cm
4.2 m to 4.5 m	22 cm	Ø 12 mm @ 15 cm	Ø 8 mm @ 20 cm

In inches:

Span L	Slab thickness h	Primary reinforcement	Secondary reinforcement
up to 10'	6"	# 3 (Ø 3/8") @ 5"	8 mm or # 3 (Ø 3/8") @ 6"
10'-1" to 12'	7"	# 4 (Ø 1/2") @ 6"	8 mm or # 3 (Ø 3/8") @ 8"
12'-1" to 14'	8"	# 4 (Ø 1/2") @ 6"	8 mm or # 3 (Ø 3/8") @ 8"
14'-1" to 15'	9"	# 4 (Ø 1/2") @ 6"	8 mm or # 3 (Ø 3/8") @ 8"

Figure 176: Tables of rebar sizes and spacing for full concrete slabs, in metric units and inches

Curing of full concrete slabs

Create little dams with mud and fill regularly with water for 2 weeks (Figure 177). Alternatively you may cover the slab with plastic sheeting which will have to be removed every day to water the slab.

Remove the slab supporting formwork only after 3 weeks.

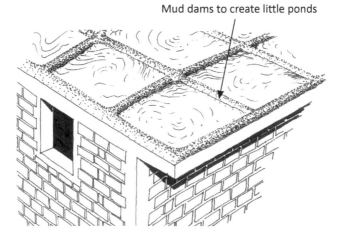

Mud dams to create little ponds

Figure 177: Keep slab wet for 2 weeks

11.3   Pan and joist slabs ('lightweight slabs')

Pan and joist slabs, also known as lightweight slabs, are made of prefabricated hollow block elements called 'pans' and joists which can be prefabricated or poured in place. The formwork for precast joists is limited to supporting timber beams, while cast in place joists need additional boards. Pan elements can be made of concrete, brick or even polystyrene (which results in a genuine lightweight slab). The upper part of such composite slabs is made of a layer of freshly poured concrete (Figure 178).

Figure 178: Left: Concrete pans with joists poured on site; Centre: Brick pans with prefab concrete joists, Right: Concrete pans with prefab concrete joists

Primary reinforcement

The primary reinforcement is placed in the joists between the concrete elements. The reinforcement consists of one or two lower rebars (number mostly due to local practice) and can contain an upper rebar as well. It is vital that these rebars are well connected with the ring-beam. This means they have to be bent into the ring-beam (Figure 179).

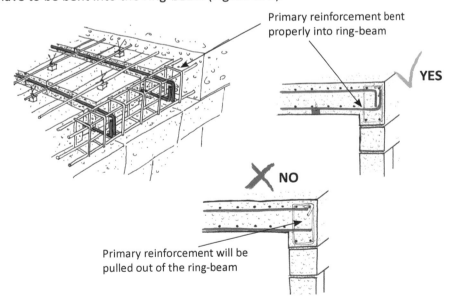

Figures 179: Proper connection of primary reinforcement with ring-beam

Steel reinforcement necessary for joists in a 20 cm (8 in.) lightweight slab

Figure 180 gives the diameters and positioning of rebars in lightweight slab joists spaced at 40 cm. For longer spans consult an engineer.

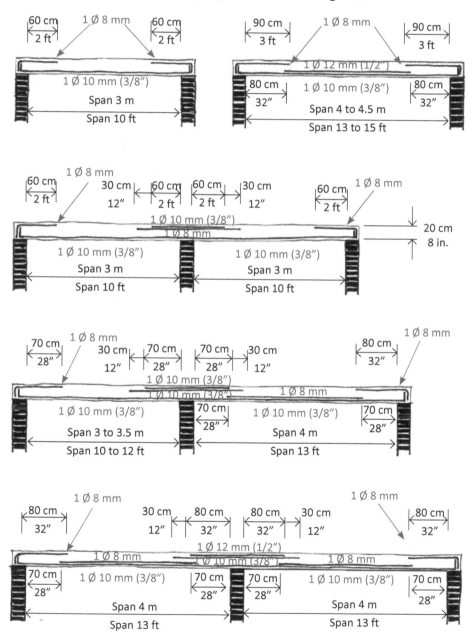

Figures 180: Steel reinforcements in lightweight slab joists

Secondary reinforcement (shrinkage or temperature reinforcement)

The temperature reinforcement bars are laid on top of the primary reinforcement and across the pan elements. Make sure to place all rebars on 3 cm (1-1/4 in.) spacers (Figure 181). If the rebars are not properly embedded in concrete they won't perform properly. The rebar ends are bent into the ring-beams (Figure 182).

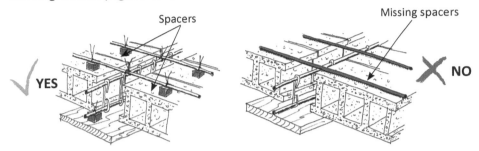

Figures 181: All rebars must be placed on spacers

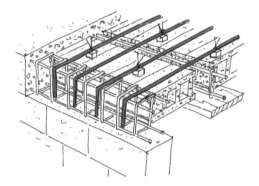

Figure 182: Rebar ends bent into the ring-beams

Pipes in lightweight slabs

Water and drainage pipes should not cross slab joists as they weaken them. Plan your project in such a way that pipes crossing joists are not necessary. If unavoidable, do not cross more than one joist.

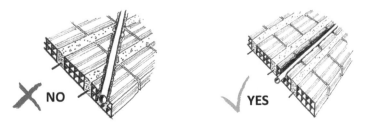

Figures 183: Don't let pipes cross the joists

If you cannot prevent pipes from crossing joists, place a double joist in the crossing area (Figure 184). Don't drill holes in the joists for light fittings (Figure 185). Make crossings as close as possible to midspan.

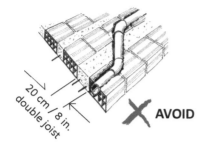

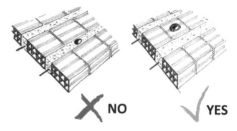

Figure 184: Use double joists where pipes must to cross

Figure 185: Do not pierce the joist but rather the pan

Pouring the concrete

Before pouring the concrete slabs, verify that the water and drainage pipes do not leak (Figure 186). Temporarily block the pipes and leave open only one end. Fill the pipe with water and verify after 4 hours that all pipe connections are dry and that there has not been any water leakage.

Figure 186: Check that all pipes are watertight

Figure 187: Wet the blocks and formwork

Water the pans and the formwork thoroughly before starting to pour the concrete (Figure 187). Place a board to walk on. Don't step on the pans and the reinforcement.

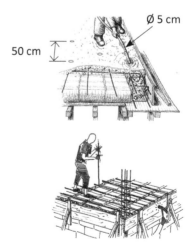

Pour the concrete and compact well, ideally with a needle vibrator. Put the needle in quickly and pull it back slowly in 5 to 15 seconds depending on the height of the concrete layer. Poke the needle in the concrete every 50 cm (10 x diameter of needle). If no needle is available, compact with a rod while hammering against the lateral formwork (Figure 188).

Figure 188: Compact the concrete well to get the air pockets out

Use a wooden or metal straightedge to smooth and level the concrete mix (Figure 189). If there are areas below the desired level fill the holes with concrete. Constantly verify that the slab surface is level.

If for any (bad) reason concrete work has to be interrupted, let the construction joint be in the first third of the slab, never in its center (Figure 190). Do make a rough sloping joint. This will make it easier for the next batch of fresh concrete to connect well with the first one. Rebar splices should be at least 60 bar diameters away from the joint.

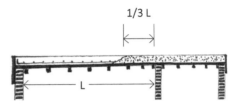

1/3 L

Figure 190: Interrupt concrete work in the first third of the slab

Figure 189: Level the concrete with a straightedge

Curing of fresh slab

Create little ponds on the slab by making small dams with sand or mud. Fill the ponds with water and refill them for at least two weeks (Figure 191). Start to cover the slab with water as soon as the concrete has hardened. Don't wait until the next day, as too much water will evaporate in the meantime. Avoid working on the fresh slab for two days.

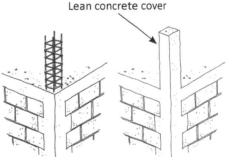

Figure 191: Keep the slab under water for two weeks

Lean concrete cover

Removal of formwork

Leave the supporting formwork for three weeks. The lateral parts of the formwork can be removed after 2 or 3 days.

Figures 192: Cast free vertical rebars in lean concrete

11.4   Embedding free vertical rebar

If you have left the vertical rebar projecting from the ring-beam and slab, cast them in lean concrete (Figure 192). This concrete will protect the rebar against rust, while being easily removable in the future.

Moreover, the additional concrete will offer a greater anchoring length for the vertical rebar so that they cannot be pulled through the ring-beam during an earthquake (Figure 193).

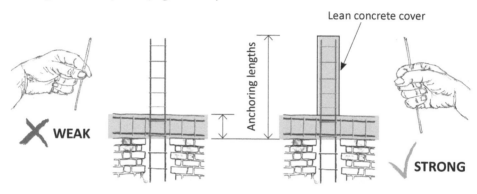

Figure 193: A longer anchoring length for the vertical rebars makes the connection stronger

The posts made from the free vertical rebars (in French: 'fers en attente' = 'waiting rebars') can be used to anchor terrace walls or act as supports for shade or lightweight roofs (Figure 194). For seismic bands follow the same rules as described above for the walls (at 1.2 m above ring-beam).

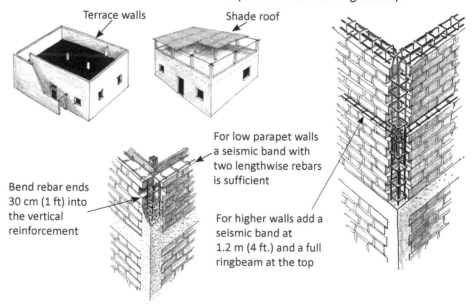

Figure 194: Use the free vertical rebars to anchor parapets, terrace walls or shade roofs

http://dx.doi.org/10.3362/9781780449906.012

CHAPTER 12 STAIRS

A beautiful stair that breaks down after some years is as bad as a solid stair that is very uncomfortable. For this reason, two sets of rules have to be observed: one related to comfort and the other related to engineering.

12.1 Comfort rules

Comfort rules ensure the functionality and safety of stairs. The following guidance is valid for stairs in family homes. For public places such as restaurants where more people will use the stairs, the rules are different. Country-specific rules and regulations may also apply.

Slope and step ratio

If a flight of stairs is too steep with steps too high and too short there is a risk of falling, which is why the inclination of a flight of stairs must follow two rules:

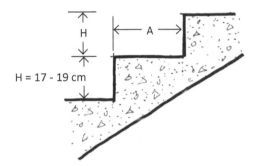

Figure 195: Step ratio

1. Step ratio: A + 2H = 61 to 65 cm (25" to 26")

2. As indicated in Figure 195, the maximum height H of a step should be between 17 and 19 cm (7" to 7 1/2")

Head clearance above stairs

Plan your staircase well before making the slab. The passage for the stairs must allow for a head clearance of 1.8 m (6 ft.) between the steps and the edge of the slab. This height will not only cause you less headache, but will facilitate the transport of furniture to the upper floor (Figure 196).

First and last step

If you cover the stairs with the same tiles as the floors (both lower and upper) then your raw steps are all of the same height. However, if the stairs are left in concrete while the floors are tiled, or if the tiles on the steps are thinner than the floor tiles, the first and last raw concrete steps must be of different heights than the rest.

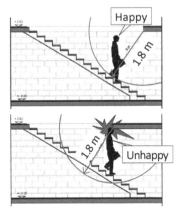

Figure 196: Head clearance 1.8 m (6 ft)

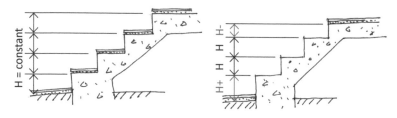

Figure 197: Problem of first and last step

12.2 Engineering rules

Figure 198 illustrates typical reinforcement details for two-span stairs.

- Diameter of all rebars: 12 mm = 1/2 in.
- Spacing of all lengthwise rebars: 15 cm = 6 in.
- Spacing of all crosswise rebars: 25 cm = 10 in.

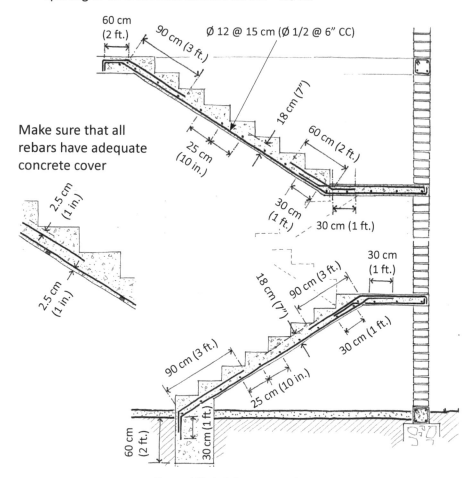

Figure 198: Reinforcement of two-span stairs

http://dx.doi.org/10.3362/9781780449906.013

CHAPTER 13   HOUSE EXTENSIONS

13.1   Vertical extensions

The recommendations and construction details given in this manual allow for a maximum extension of one floor above the ground floor.

You can add one floor only if the free vertical rebars (left to connect later to the upper vertical ties) have been embedded in lean concrete. Never use free vertical rebars which have been exposed and have become rusty (Figure 199).

Vertical extension only if the free rebars have been protected with lean concrete

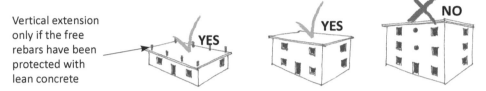

Figure 199: Maximum vertical extension: ground floor plus one

The rules for the construction of the upper floor are the same as for the ground floor, including the creation of seismic bands and reinforcement of openings.

Vertical ties have to be continuous from the ground floor to the upper floor. The overlapping length of the vertical rebars is 60 times their diameter (Figure 200). The number of stirrups is doubled in the overlapping area. leading to a spacing of 10 cm.

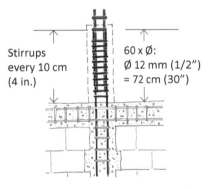

Stirrups every 10 cm (4 in.)

60 x Ø: Ø 12 mm (1/2") = 72 cm (30")

Figure 200: Overlapping length of vertical ties is 60 x Ø

13.2   Horizontal extensions

As a general rule it is better to detach buildings rather than to add extensions to an existing house, particularly if the shape or the height of the two buildings are different.

As shown in Figure 201, the minimum theoretical distance between two buildings is 1/100th of their height (height: 3 m = distance of 3 cm). However, in reality it is impossible to have a clean empty joint, without mortar falling into and blocking it. This is why a minimum joint of 15 cm (6 in.) is recommended (Figure 202). The joint can be closed laterally with bricks placed against the walls (not in-between) to allow for free movement.

Figure 201: Theoretical minimum distance between buildings

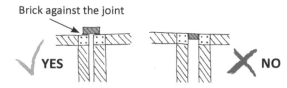

Figure 202: Closing the gap with bricks at the outside

If new rooms have to be solidly attached to the original house or if you want to make a horizontal extension both at the ground and upper floor, make sure that the overall shape of the complete building remains simple and that the proportion between long and short façade remains 1 : 3 (Figure 203). Make sure to create solid confined masonry bracing walls in the extension.

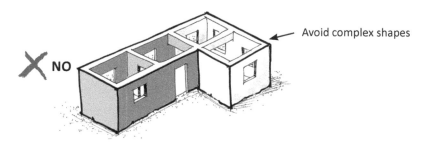

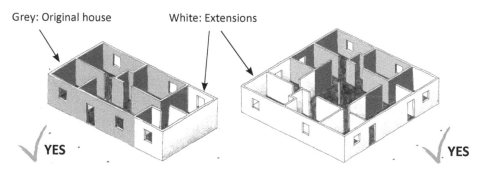

Figure 203: Maintain a regular shape when adding rooms

In order to connect a new structure solidly to the old one, observe the following procedure, illustrated in Figure 204:

1. Open all joints towards the extension, and free the rebars.
2. Prepare the foundations for the extension at the same level as the original one.

118

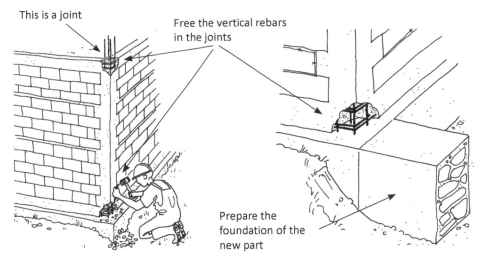

This is a joint

Free the vertical rebars in the joints

Prepare the foundation of the new part

Figure 204: Opening up existing corner connections

3. Prepare Ø 8 mm (3/8 in.) rebars with hooks at the ends that hook around the existing building reinforcing (Figure 205).

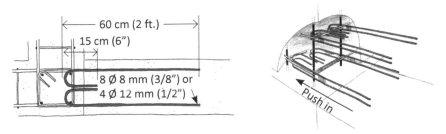

60 cm (2 ft.)

15 cm (6")

8 Ø 8 mm (3/8") or 4 Ø 12 mm (1/2")

Push in

Figure 205: Add hooked connection bars

4. Hook eight of these rebars around the original vertical rebars of each knot and push them back into the knot as much as possible. If you don't have Ø 8 mm bars, use 4 Ø 12 mm (1/2") rebars .

5. Place the horizontal and vertical tie reinforcements of the extension (Figure 206).

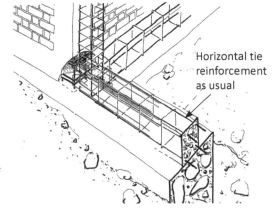

Horizontal tie reinforcement as usual

Figure 206: Integrate the connection bars into the new plinth beams

119

6. Make the walls and cast the concrete ties around them. You will end up with a double tie structure where the extension meets the original building (Figure 207).

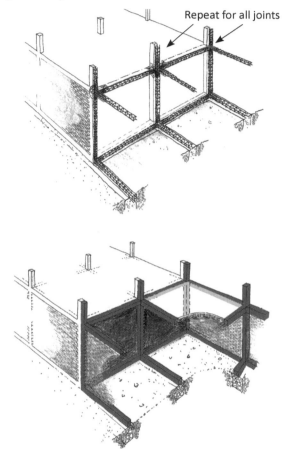

Figure 207: The added structure is doubled on the side of the original building

Part III:
Additional issues

http://dx.doi.org/10.3362/9781780449906.014

CHAPTER 14   THE SHOP WINDOW PROBLEM

Wide shop window fronts reduce the stability of a building during an earthquake. By adding confined stabilizing walls the rigidity of the shop front can be increased. Unfortunately these stabilizing walls will reduce the length of the shop windows. However, with some smart design solutions such as placing the glass front further back, the size of the shop windows can be maintained.

A further issue needs to be addressed: shops often consist of deep rooms with no intermediate cross walls (Figure 208). A similar layout does weaken the building. To increase the stability of the long walls, intermediate transverse walls should be added.

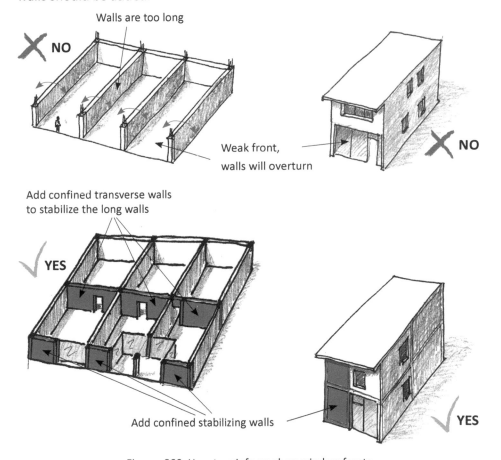

Figures 208: How to reinforce shop window fronts

CHAPTER 15   RETAINING WALLS IN STONE MASONRY

15.1   General aspects

Houses are often built on slopes. In order to create a level surface for the building, retaining walls are needed. They can be built in concrete, under the supervision of a qualified engineer. In poor countries it is usually less expensive to build them with local stone. To make them safe, observe the following rules and limit their height to 2.5 m (8' - 4".).

Plan the site for your house in such a way that the house can be placed on firm original soil (Figure 209). The backfill behind retaining walls will not be as solid as the soil which has not been touched.

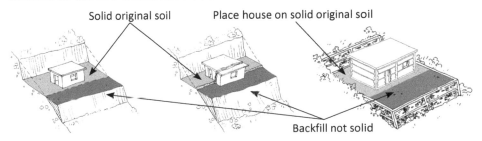

Figure 209: Build house on firm soil, even with a retaining wall

The house should not be built on top of or against a retaining wall (Figure 210). It is wise to keep a passage way behind the house which will allow rain water to drain off sideways (instead of running into your house).

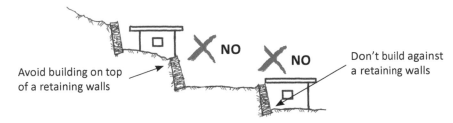

Figure 210: Keep away from retaining walls

As illustrated in Figure 211, the distance between the house front and the lower retaining wall has to be at least as wide as the lower retaining wall is high (H = H).

Keep the passage way behind the house large enough for a person to pass through for cleaning. 60 cm (2 ft.) is a good minimum distance.

Avoid high retaining walls. If needed, make several lower walls and space them at the same distance as the lower retaining wall height (h = h). See Figure 212.

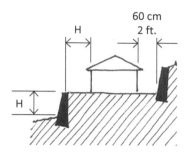

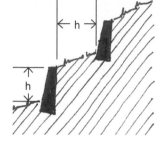

Figure 211: Distances from house to walls   Figure 212: Distances between retaining walls

15.2 Construction rules

There are a few basic rules one should follow to build a safe retaining wall in stone masonry:

1. Maximum height of a stone retaining wall made without the assistance of a qualified engineer is 2.5 m (8'4") and the top of the wall should be at least 60 cm (2 ft.) wide (Figure 213).

2. The front of the retaining wall has to be inclined towards the slope in a ratio of 5 to 1 (Figure 214). This means, for every 50 cm up lean back 10 cm (or: for every 50 in. up, lean back 10 in.)

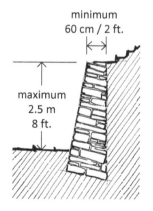

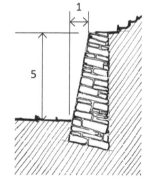

Figure 213: Height and thickness   Figure 214: Inclination ratio 1 : 5

3. The width of the foundation should be equal to half of the free height of the wall. The bottom of the foundation should be inclined towards the slope (Figure 215).

4. The depth of the foundation depends on the type of soil and the type of climate. In areas where the soil freezes in winter, the foundation must start below the freezing depth. Find out about the freezing depth in your locality (Figure 216).

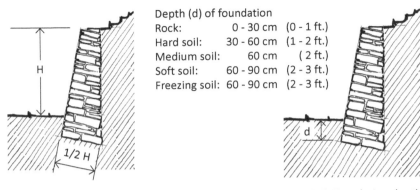

Depth (d) of foundation
Rock: 0 - 30 cm (0 - 1 ft.)
Hard soil: 30 - 60 cm (1 - 2 ft.)
Medium soil: 60 cm   (2 ft.)
Soft soil: 60 - 90 cm (2 - 3 ft.)
Freezing soil: 60 - 90 cm (2 - 3 ft.)

Figure 215: Foundation width

Figure 216: Foundation depth

5. Stones must be placed flat, not on their side (Figure 217). Use bigger rather than smaller stones. Place through-stones at regular intervals that go through the whole width of the wall.

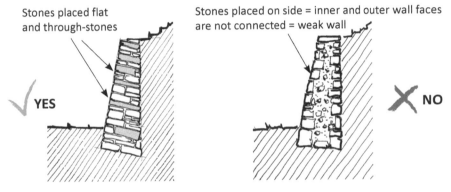

Stones placed flat and through-stones

Stones placed on side = inner and outer wall faces are not connected = weak wall

YES

NO

Figures 217: Stones placed flat, not on side

6. Stone layers must be inclined at a right angle to the front (Figure 218). This way they won't be pushed out by soil pressure.

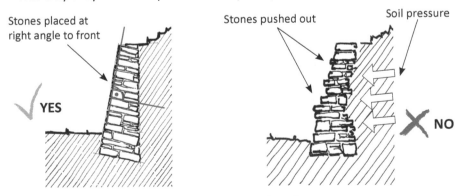

Stones placed at right angle to front

Stones pushed out

Soil pressure

YES

NO

Figure 218: Stones at right angle to front

7. In order to reinforce the whole wall, you may add reinforced concrete bands (Figure 219).

8. Don't forget the drainage pipes every 1.5 m (5 ft.) in the lower part of the wall, and at mid-height if the wall is 2.5 m (8' - 4") high (Figure 220).

Reinforced concrete bands

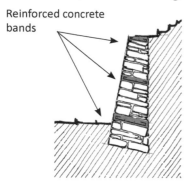

Drainage pipes

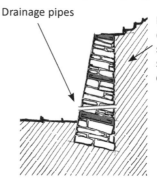

Backfill with draining material such as gravel or stone, not with clay or clay soil

Figure 219: Adding reinforced concrete bands

Figure 220: Don't forget drainage pipes

15.3   Examples

Some examples of retaining walls reinforced with concrete bands are given below in Figure 221. All dimensions have been simplified and rounded for easier use.

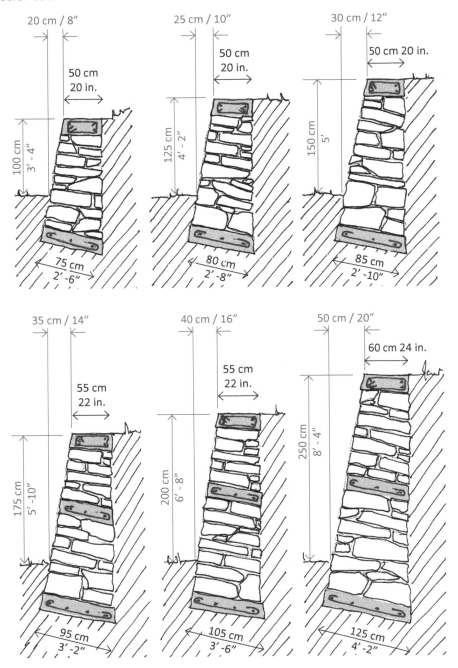

Figures 221: Retaining wall proportions from 1 to 2.5 m (3' - 4" to 8' - 4")

http://dx.doi.org/10.3362/9781780449906.016

CHAPTER 16   PITCHED LIGHTWEIGHT ROOFS

In certain climatic, economic or cultural environments people prefer pitched roofs on their houses. The following pages provide some basic instructions to build economic but solid roofs, without going into all the details of carpentry. Additional hints are given to improve the hurricane resistance of lightweight roofs.

16.1   Roof shapes

Pitched roofs can be classified as 'gable roofs' and 'hip (or 'hipped') roofs', each with its advantages and disadvantages. Gable roofs are easier to make, but hipped roofs can be more hurricane-resistant if built properly (Figure 222).

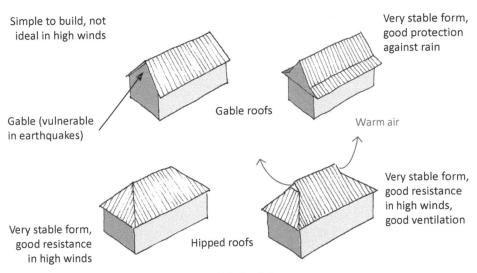

Simple to build, not ideal in high winds

Very stable form, good protection against rain

Gable (vulnerable in earthquakes)

Gable roofs

Warm air

Very stable form, good resistance in high winds

Hipped roofs

Very stable form, good resistance in high winds, good ventilation

Figures 222: Roof shapes

16.2   Hurricane resistance

A hurricane-resistant house must not present any opening or big eaves where the winds can attack. Windows must be shuttered when a storm approaches. Porticos should have independent roofs which can get blown off without damaging the main roof. If a house is placed on posts, the void under the house must be closed off to avoid wind penetration (Figure 223).

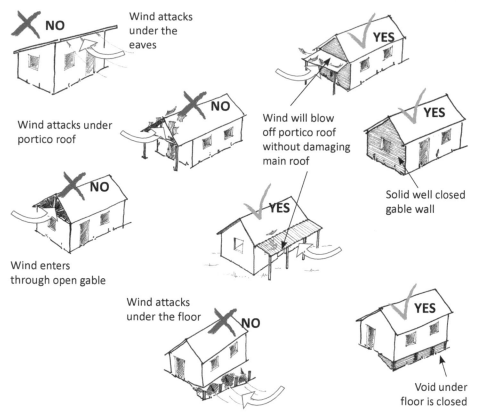

Wind attacks under the eaves

NO

YES

Wind attacks under portico roof

NO

Wind will blow off portico roof without damaging main roof

YES

Solid well closed gable wall

NO

Wind enters through open gable

YES

Wind attacks under the floor

NO

YES

Void under floor is closed

Figures 223: Do's and dont's against hurricanes

16.3 Roof structure

Unreinforced gable walls are inherently weak when it comes to earthquakes. They tend to collapse easily (Figure 224). If the roof structure is placed on the gable walls and is not properly braced, the whole roof will collapse during a quake. Therefore, gable walls in masonry have to be confined like all other walls of the house, and roof structures have to be braced.

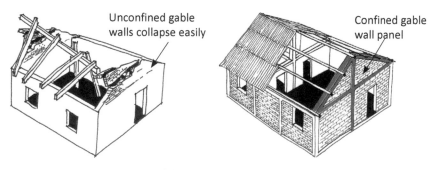

Unconfined gable walls collapse easily

Confined gable wall panel

Figure 224: Gable walls have to be confined

Slabs:

1. Use straight vertical posts. Bent or curved sticks will bend and can break under the weight of the slab.
2. Use good boards with straight edges to avoid leaking.
3. For minor repairs to close gaps between boards you may use mud.
4. Never place the poles on stacks of bricks. They might fall over, leading to the collapse of the entire formwork.

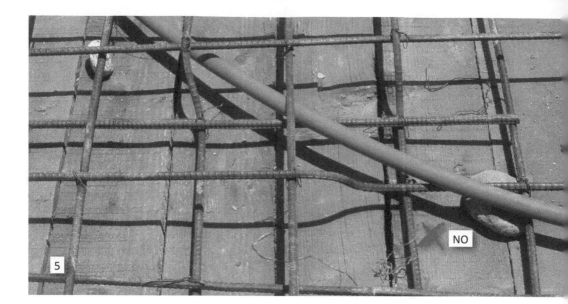

5 NO

6 NO

Slabs:

5. Don't use stones but proper spacers. They will slip away when the concrete is poured.
6. Don't fill the gaps with paper and junk, but make proper formwork.
7. Rebars have fallen on the formwork because of no or bad spacers. Exposed rebars will rust quickly and loose their strength.

7 NO

8

Slabs:

8. Joist and pan slab, with joists poured on site. Use this method only if concrete blocks are of good quality.
9. Keep slab under water for two weeks. Free vertical rebars are too short for future use, and should be embedded in lean concrete.
10. Free vertical rebars too short. Never place a pipe in a vertical tie.
11. Free vertical rebars too short for future use.

9

10

11

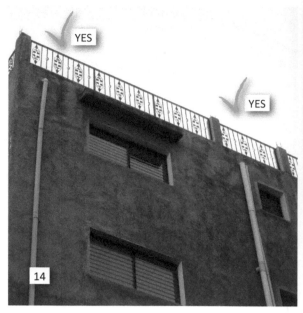

Slabs:

13.   Free vertical rebars too short for future use.

12., 14. Free vertical rebars cast in concrete can be used to support a sun or rain roof or can be part of a balustrade.

15

16

17

18

Retaining walls:

15., 18. Never build a retaining wall with stones placed on their side and rubble at the core. Stones must be placed flat and throughstones must reach from inside to outside.

16. Never use rubble stone to build a retaining wall.

17. Front must be inclined at a 1:5 ratio. Reinforced concrete bands at regular intervals will strengthen the wall.

19

YES

20

YES

Retaining walls:

19. Stones must be pitched at a right angle to the front.
20. Drainage holes are very important to reduce the water pressure behind the wall.
20., 21. Limit the height of a retaining wall to 2.5 m. Avoid buidling directly on top of a retaining wall.

AVOID

NO

21

Pitched roofs and hurricanes:

22. Gable walls must be confined like any other wall panel.
23. Place roof anchors in the ring beam.
24., 25. Independent lateral or front roofs may be damaged by a hurricane without jeopardizing the main roof.

Pitched roofs and hurricanes:

26. Avoid continuous roof over both the main body and the veranda. A hurricane can demolish the entire roof.
27. Use timber boards or metal nail plates to place enough nails in the joints.
28. Use solid steel straps to tie the roof structure.

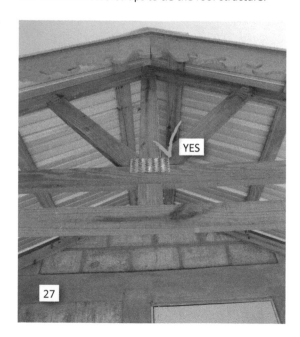

The reinforcement of the gable ties follows the same rules as those for horizontal beams. Take care to bring the lower rebars towards the outside, together with the ends of the vertical rebars (Figures 225 and 226).

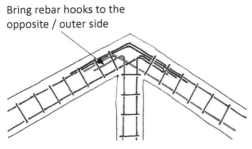

Bring rebar hooks to the opposite / outer side

Figures 225: Reinforcement of gable ties

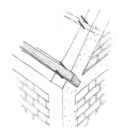

Figures 226: Placing anchors in the gable ties

With a solid confined gable wall, the ridge beam and purlins can be placed directly on these concrete elements. Fewer trusses are needed. Their number will depend on the length of the room and the height of the ridge beam and purlins. The beams and purlins can be fixed to the concrete structure with rebar anchors placed earlier in the concrete ties (Figure 227).

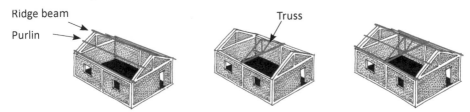

Ridge beam

Purlin

Truss

Figure 227: Purlin size and number of trusses depend on the length of the roof

The construction of solid roofs without confined gable walls is also possible. More trusses are needed and wind braces are of utmost importance (Figure 228). Wind braces should also be put in hipped roofs. Hip rafters alone can only act as braces if they are very solidly attached both to the ridge and the top plates (which in turn must be well linked in the corners).

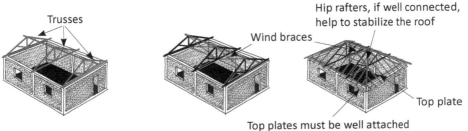

Trusses

Hip rafters, if well connected, help to stabilize the roof

Wind braces

Top plate

Top plates must be well attached to each other, all around

Figure 228: Roof trusses and wind braces

The type of roof truss depends on the span (Figure 229). When preparing the trusses, always use many nails. Four or six nails are usually not enough. Adding nail boards (or metal nail plates) makes the trusses stronger. Don't place the trusses more than 3 m (10 ft.) apart.

Only for short spans

4 - 6 nails are not enough!

Good for medium spans and larger eaves

Strong truss for wider spans

Nail boards

Figure 229: Different types of roof trusses

Braces must be well attached to the roof structure (Figure 230). Use sufficient nails and straps. If you're not sure about the quality of the connections (e.g. insufficient number of nails) add more cross braces further down the roof.

Diagonal bracing in the vertical plane

Diagonal bracing in the roof plane

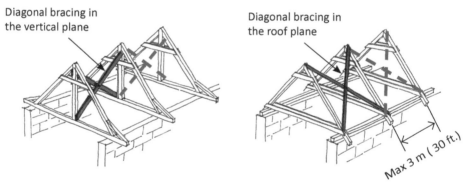

Max 3 m (30 ft.)

Figure 230: Bracing can be done both ways

16.4 Connections

The strongest roof structure is of little use if it is not well attached to the rest of the house. Anchor bolts or rebars embedded in the concrete elements of the walls do offer a solid connection for the roof structure.

Proper U-shaped anchor bands with a minimum thickness of 3 mm (11 gauge) are the best solution.

If proper anchor bands cannot be found, Ø 12 mm (1/2") rebars can be used. It is critical that the rebars be nailed down with long solid nails, otherwise the timber elements could move in an earthquake or be raised by the wind (Figure 231).

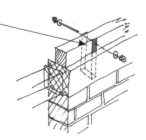

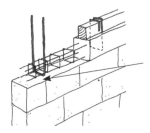
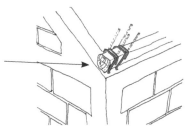

Embedded Ø 12 mm (1/2") rebars bent around main reinforcement and timber elements and nailed down with several 12 cm (5"/40d) nails

Figure 231: Anchoring the roof structure with anchor bands or rebars embedded in the concrete ties of the wall

Do not use the free rebars of the main vertical ties to tie down the timber pieces. Rebars which have already been bent cannot be used for future vertical extensions.

When using straps, make sure that they are made of a solid metal sheet Figure 232). Straps should have a minimum width of 25 mm (1 in.) and minimum thickness of 1.2 mm (18 gauge). If such metal cannot be found, you may use straps cut out of high quality CGI (corrugated galvanised iron) sheets with a minimum thickness of 0.7 mm (22 gauge). Make these thinner straps slightly larger (40 mm / 1.5").

Don't use straps cut out from thin 0.4 mm (28 gauge) CGI roofing sheets or from cans.

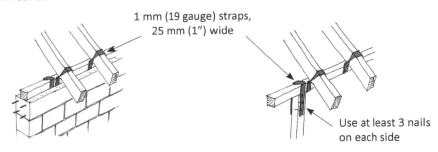

1 mm (19 gauge) straps, 25 mm (1") wide

Use at least 3 nails on each side

Figure 232: Metal straps must be solid or they are of no use

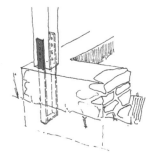

Finally, don't forget to attach all veranda posts to the foundation or a hurricane will blow them away together with the veranda roof (Figure 233).

Figure 233: Anchor the veranda posts too

16.5 Covering

The roof may be covered with clay tiles or corrugated galvanised iron (CGI) sheets. If tiles are used, the roof structure will likely be sturdier because of the weight of the tiles. Also, it is better to use a type of tile which can be nailed down.

CGI sheets should be of a good and long lasting quality. The usual 0.4 mm (28 gauge) sheets are too thin and will rust away in a few years. The recommended thickness for CGI sheets, though difficult to find, is 0.7 mm (22 gauge). They will last for a long time.

The single CGI sheets should be placed 'against the wind'. This way the resulting overlap of the sheets will not be exposed to the aggressive power of the wind (Figure 234). The overlap should go over two crests. To avoid water infiltration, nails must go through the crests, not through the troughs (Figure 235). A rubber washer should be added to the nails to increase water tightness. If not available locally, it can easily be made with bits of tyre inner tubes.

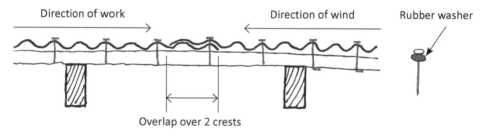

Figure 234: Horizontal overlap for CGI cover

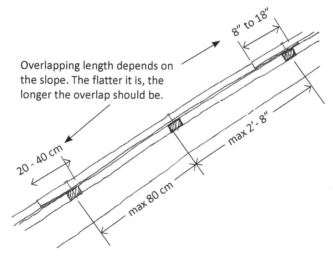

Figure 235: Slope overlap for CGI cover

144

http://dx.doi.org/10.3362/9781780449906.017

CHAPTER 17   QUALITY ASSURANCE

Quality assurance is the monitoring of materials and inspection of workmanship that are critical to the integrity of the building structure. It helps to ensure compliance with the approved plans and specifications and relevant codes, ordinances, and guidelines. This not only includes verification of material strengths and placement inspection but also verification that proper construction techniques are being followed.

Many regions where confined masonry construction is common have quality assurance provisions already included as part of the governing building code. Unfortunately, there are other regions where quality assurance is either not part of the building code or are not fully enforced. There are also some building codes that may waive the code prescribed quality assurance requirements for single family houses and the non-engineered buildings that are covered in these guidelines. However, it is still important that the builders get it right, even if there is no formal quality assurance programme enforced by the local government agency.

To that end, it is recommended that a quality assurance plan be developed and enforced. The building owner should lead the effort to establish and monitor the plan since he/she will gain the most benefit by having such a programme in place. The owner should hire the quality assurance inspectors and meet with them periodically during construction to verify that the construction and inspection is in accordance with the quality level that the owner desires. Preferably the inspectors should be from an inspection agency that specializes in inspecting the type of construction being performed. However, it may be reasonable for the building designer or the building owners themselves to perform at least some of these inspections, provided that they are properly trained in how to perform these inspections.

It is important that the persons involved in inspection and quality assurance testing be independent from the builder in order to avoid a direct conflict of interest. The intent of the inspection and testing is to verify the quality of the builder's work, and thus the builder should not be in a position of performing or directing the inspections. The builder may have a separate in-house quality control programme. While such a programme can be beneficial to establishing a level of construction quality it should not take the place of an independent quality assurance plan.

Many quality assurance tasks have been described in other sections of these guidelines. Others have been described in construction guideline documents developed by various agencies, some of which are available for download at the Confined Masonry Network website, www.confinedmasonry.org.

An easy way to create a quality assurance plan and monitor the compliance to it is by using checklists. A sample checklist has been provided in this guideline for reference. This checklist includes a representative sample of the types of inspections that can be performed on a confined masonry project. *It is not a comprehensive list.* Items can be added and deleted to suit the needs of a particular project, or to comply with local building regulations, or for other reasons. The non-government organization Build Change has developed comprehensive checklists for use in the countries where they work. If the reader would like further information or a copy of these checklists then he/she is encouraged to contact Build Change at www.buildchange.org.

SITE and SOIL CONDITIONS CHECKLIST

Homeowner: _____ Address: _____

Designer: _____

Builder: _____

		Planned?	Date	Comments	Built?	Date	Comments
1	**DO NOT BUILD ON STEEP SLOPE**	**Planned?**	**Date**	**Comments**	**Built?**	**Date**	**Comments**
a	Level Site	Yes / No			Yes / No		
2	**SETBACKS from STEEP SLOPES**	**Planned?**	**Date**	**Comments**	**Built?**	**Date**	**Comments**
a	Sufficient distance behind house to slope	Yes / No			Yes / No		
b	Sufficient distance in front of house to slope	Yes / No			Yes / No		
c	No loose debris, falling soil or rock close to house	Yes / No			Yes / No		
d	No existing building close to house upslope of site	Yes / No			Yes / No		
3	**IDENTIFY SEISMIC HAZARD**	**Planned?**	**Date**	**Comments**	**Built?**	**Date**	**Comments**
a	Site is located in an area of medium or high seismicity?	Yes / No			Yes / No		
b	Site is located in an area of very high seismicity?	Yes / No			Yes / No		
4	**SETBACKS from RIVER and DRAINAGE**	**Planned?**	**Date**	**Comments**	**Built?**	**Date**	**Comments**
a	Sufficient distance from riverbed or drainage channel	Yes / No			Yes / No		
5	**IDENTIFY SOIL TYPE & SCREEN FOR HAZARDOUS SOILS**	**Planned?**	**Date**	**Comments**	**Built?**	**Date**	**Comments**
a	Soil is rock	Yes / No			Yes / No		
b	Soil is compact gravels and compact sands	Yes / No			Yes / No		
c	Soil is non-consolidated sand, silt, soft clay) and…	Yes / No			Yes / No		
d	If sandy soil, there exists no risk of liquefaction (asses water table location)	Yes / No / NA			Yes / No / NA		
e	Soil is not expansive clay	Yes / No			Yes / No		
6	**SCREENING FOR OBSTACLES ON SITE**	**Planned?**	**Date**	**Comments**	**Built?**	**Date**	**Comments**
a	There are no large obstacles that need to be removed (trees, existing walls etc)	Yes / No			Yes / No		
b	The site is not covered in fill material	Yes / No			Yes / No		
7	**SETBACKS FROM ROADS and BUILDINGS**	**Planned?**	**Date**	**Comments**	**Built?**	**Date**	**Comments**
a	Setback at least 2 - 3 m from road or front boundary	Yes / No			Yes / No		
b	Setback at least 2 - 3 m from side boundary	Yes / No			Yes / No		
c	Rainwater can flow into drainage	Yes / No			Yes / No		
d	Building constructed at least 2 - 3m behind the fence	Yes / No / NA			Yes / No / NA		

Homeowner Signature: _____ Date: _____

Designer Signature: _____ Date: _____

Builder Signature: _____ Date: _____

CONFIGURATION CHECKLIST FOR SINGLE AND 2 STOREY BUILDINGS

Homeowner: _____

Designer: _____ Address: _____

Builder: _____

		Planned?	Date	Comments	Built?	Date	Comments
1	**PLAN**						
a	Length to width ratio equal to 4 or less	Yes / No			Yes / No		
b	Separate irregular shaped buildings (L,U,E)	Yes / No / NA			Yes / No / NA		
2	**ELEVATION**	**Planned?**	**Date**	**Comments**	**Built?**	**Date**	**Comments**
a	Building has not more than 2 stories	Yes / No			Yes / No		
b	Maximum height of ground floor walls = 3 m	Yes / No			Yes / No		
c	Wall height to thickness ratio no more than 22	Yes / No			Yes / No		
3	**MINIMUM SHEAR WALL DENSITY**	**Complies?**	**Date**	**Comments**	**Built?**	**Date**	**Comments**
a	Only include properly confined walls longer than 2/3 its height in shear wall density calculation	Yes / No			Yes / No		
b	Shear wall density complies with guidelines	Yes / No			Yes / No		
4	**SHEAR WALL LOCATION**	**Planned?**	**Date**	**Comments**	**Built?**	**Date**	**Comments**
a	At least two lines of shear walls in X direction	Yes / No			Yes / No		
b	At least two lines of shear wallls in Y direction	Yes / No			Yes / No		
c	Shear walls are symmetrically placed	Yes / No			Yes / No		
d	Shear walls are as far as possible from one another	Yes / No			Yes / No		
e	Shear walls are on exterior of building	Yes / No			Yes / No		
f	Spacing of perpendicular or cross walls does not exceed guidelines	Yes / No			Yes / No		
5	**TIE COLUMN LOCATION**	**Planned?**	**Date**	**Comments**	**Built?**	**Date**	**Comments**
a	Every corner (L)	Yes / No			Yes / No		
b	Every wall intersection (T)	Yes / No / NA			Yes / No / NA		
c	Every change in the direction of the wall	Yes / No			Yes / No		
d	At both ends of every shear wall	Yes / No			Yes / No		
6	**TIE BEAM LOCATION**	**Planned?**	**Date**	**Comments**	**Built?**	**Date**	**Comments**
a	At the foundation (plinth beam)	Yes / No			Yes / No		
b	At the roof level (ring beam)	Yes / No			Yes / No		
c	Intermediate ring beam at sill level	Yes / No			Yes / No		
d	Seismic bands	Yes / No / NA			Yes / No / NA		

		Planned?	Date	Comments	Built?	Date	Comments
7	**OPENING SIZE**						
a	Maximum 1/2 length between to crosswalls	Yes / No			Yes / No		
b	Openings positioned directly under tie beam or horizontal band	Yes / No			Yes / No		
c	Doors reinforced on both sides with tie beam or column or vertical band	Yes / No			Yes / No		
d	Windows reinforced on both sides with tie column or vertical band	Yes / No / NA			Yes / No / NA		
8	**TWO STORY CONFIGURATION RULES**	Planned?	Date	Comments	Built?	Date	Comments
a	No soft story (open ground floor)	Yes / No			Yes / No		
b	No slab roof overhangs	Yes / No / NA			Yes / No / NA		
c	Columns are continuous both floors	Yes / No / NA			Yes / No / NA		
d	Shear walls line up vertically	Yes / No / NA			Yes / No / NA		
e	Openings line up vertically	Yes / No / NA			Yes / No / NA		
f	No second story over a porch	Yes / No / NA			Yes / No / NA		
g	Second story column bar lap detailed	Yes / No / NA			Yes / No / NA		

Homeowner Signature: _____ Date: _____

Designer Signature: _____ Date: _____

Builder Signature: _____ Date: _____

MATERIALS QUALITY CHECKLIST

Homeowner: _____   Address: _____

Designer: _____

Builder: _____

		Planned?	Date	Comments	Built?	Date	Comments
1	**WATER, SAND and AGGREGATES**						
a	Use clean water (not salty)	Yes / No			Yes / No		
b	Use clean river sand	Yes / No			Yes / No		
c	Use crushed/angular gravel for concrete	Yes / No			Yes / No		
d	Maximum gravel size 20 mm for concrete	Yes / No			Yes / No		
2	**CEMENT**						
a	Use Portland cement	Yes / No			Yes / No		
b	Store off the ground and out of rain	Yes / No			Yes / No		
3	**REINFORCING STEEL**						
a	Verify Grade 60 marking (minimum) on bars	Yes / No			Yes / No		
b	Use deformed (ribbed) bars	Yes / No			Yes / No		
c	Do not use rusty or recycled bars for longitudinal bars	Yes / No			Yes / No		
d	Use 10 mm ø minimum bars for tie columns	Yes / No			Yes / No		
e	Use 10 mm ø minimum bars for tie beams	Yes / No			Yes / No		
f	Use 6 mm ø minimum bars for stirrups	Yes / No			Yes / No		
g	Cut column steel long enough for overlap ≥ ø 50	Yes / No			Yes / No		
h	Store off the ground and out of rain	Yes / No			Yes / No		
4	**BURNT CLAY BRICKS AND CONCRETE BLOCKS**						
a	Compressive strength of blocks meets design requirements	Yes / No			Yes / No		
b	Concrete block width equal to or greater than 15 cm	Yes / No / NA			Yes / No / NA		
c	Clay brick web thickness equal to or greater than 15 cm	Yes / No / NA			Yes / No / NA		
d	Block voids are less than 50 percent of the block surface	Yes / No			Yes / No		
e	No cracks or chips or partial blocks	Yes / No			Yes / No		

Homeowner Signature: _____   Date: _____

Engineer Signature: _____   Date: _____

Builder Signature: _____   Date: _____

REINFORCED CONCRETE CHECKLIST

Homeowner: _____

Designer: _____ Address: _____

Builder: _____

1	TIE-BEAMS	Planned?	Date	Comments	Built?	Date	Comments
a	25 cm high, width matching tie columns & equal or greater than masonry wall width	Yes / No			Yes / No		
b	**Longitudinal Bars**						
1	Four 10 mm longitudinal bars	Yes / No			Yes / No		
2	Minimum Strength = Grade 60	Yes / No			Yes / No		
3	Longitudinal bars ribbed or deformed	Yes / No			Yes / No		
c	*Stirrups*						
1	6 mm closed stirrups	Yes / No			Yes / No		
2	Stirrup hooks bent at 135 degrees	Yes / No			Yes / No		
3	Hook length for stirrup minimum 6 cm	Yes / No			Yes / No		
4	Cover over steel minimum 2.5 cm	Yes / No			Yes / No		
2	**TIE-COLUMNS**	**Planned?**	**Date**	**Comments**	**Built?**	**Date**	**Comments**
a	*Longitudinal Bars*						
1	Minimum section 150 mm by 150 mm	Yes / No			Yes / No		
2	Tie columns used at locations per configuration	Yes / No			Yes / No		
3	Four 10 mm deformed (ribbed) or 12 mm smooth bars	Yes / No			Yes / No		
b	*Column Ties*						
1	6 mm closed ties	Yes / No			Yes / No		
2	Stirrup hooks bent at 135 degrees	Yes / No			Yes / No		
3	Hook length for stirrup minimum 6 cm	Yes / No			Yes / No		
4	Cover over steel minimum 2.5 cm	Yes / No			Yes / No		
3	**HORIZONTAL BANDS**	**Planned?**	**Date**	**Comments**	**Built?**	**Date**	**Comments**
a	Directly above and below window level	Yes / No / NA			Yes / No / NA		
b	7.5 - 10 cm high, match width of wall	Yes / No / NA			Yes / No / NA		
c	Use two #3 (10 mm) bars	Yes / No / NA			Yes / No / NA		
d	Use #2 (6 mm) stirrups spaced at 20 cm	Yes / No / NA			Yes / No / NA		
e	Tie reinforcement into tie columns with hooks with 30 cm legs	Yes / No / NA			Yes / No / NA		
f	Minimum lap length = 50ϕ	Yes / No / NA			Yes / No / NA		
g	Pour the beam in one day	Yes / No			Yes / No		

	4 OPENING REINFORCEMENT and VERTICAL BANDS	Planned?	Date	Comments	Built?	Date	Comments
a	Form column either side of openings	Yes / No			Yes / No		
b	Use 2 #3 (10 mm) vertical bars	Yes / No			Yes / No		
c	Use #2 (6 mm) stirrups at 20 cm spacing	Yes / No			Yes / No		
d	Tie reinforcement for doors into foundation beam with hooks with 30 cm legs	Yes / No			Yes / No		
e	Tie reinforcement for windows into horizontal bands with ahooks	Yes / No			Yes / No		
f	Tie reinforcement into ring beam with hooks with 30 cm legs	Yes / No			Yes / No		
g	Minimum lap length = 50∅	Yes / No			Yes / No		
h	Reinforce concrete lintel according to standard detail	Yes / No / NA			Yes / No / NA		
	5 BAR ASSEMBLY	Planned?	Date	Comments	Built?	Date	Comments
	BEAM STIRRUPS and COLUMN TIES						
a							
1	Stirrup spacing maximum 20 cm	Yes / No			Yes / No		
2	Stirrups closely spaced (10 cm) within 1/6 the wall height of all beam-column joints	Yes / No			Yes / No		
3	Stirrup hooks rotated	Yes / No			Yes / No		
4	Stirrups tied to longitundinal bars with binding wire	Yes / No			Yes / No		
b	**_JOINT DETAILING_**						
1	Minimum lap length = 50∅	Yes / No			Yes / No		
2	Bars bent at corners and T-junctions per one of the allowable options	Yes / No			Yes / No		
3	All bent bars at corners and T-junctions bent at 90 degrees	Yes / No			Yes / No		
4	Laps tied with binding wire	Yes / No			Yes / No		
	6 FORMWORK and CONCRETE SPACERS	Planned?	Date	Comments	Built?	Date	Comments
a	Formwork is good quality (clean, not warped, no cracks, minimal number of knots)	Yes / No			Yes / No		
b	Space between steel and formwork = 3 cm	Yes / No			Yes / No		
c	Use concrete spacers as required to maintain cover	Yes / No			Yes / No		
d	Maximum size for concrete spacer is 3 cm x 3 cm x 3 cm	Yes / No			Yes / No		
e	Use binding wire in concrete spacer	Yes / No			Yes / No		
f	Check that formwork for beams is level	Yes / No			Yes / No		
g	Check that formwork for columns is plumb	Yes / No			Yes / No		
h	Check that formwork has no cracks	Yes / No			Yes / No		
	7 CONCRETE MIXING	Planned?	Date	Comments	Built?	Date	Comments
a	Use 1:2:4 or 1.5:2:3 mix	Yes / No			Yes / No		
b	Use crushed, angular gravel	Yes / No			Yes / No		
c	Use gravel with size less than 2 cm (3/4 in)	Yes / No			Yes / No		
d	Use gravel with size less than 2 cm (3/4 in)	Yes / No			Yes / No		
e	Use clean water (not salty or muddy)	Yes / No			Yes / No		

		Planned?	Date	Comments	Built?	Date	Comments
f	Use Portland Cement	Yes / No			Yes / No		
g	Mix on a clean concrete or asphalt surface, not on dirt	Yes / No			Yes / No		
h	Use a mechanical mixer if possible	Yes / No			Yes / No		
i	Batch out gravel, then sand, then cement	Yes / No			Yes / No		
j	Mix gravel, sand, and cement dry, then add water	Yes / No			Yes / No		
k	Turn over 3 times or until color is uniform	Yes / No			Yes / No		
l	Do not use too much water! Add water slowly	Yes / No			Yes / No		
m	Use slump test or hand test for water content	Yes / No			Yes / No		
8	**CONCRETE POURING and CURING**	**Planned?**	**Date**	**Comments**	**Built?**	**Date**	**Comments**
a	Clean out the bottom of tie-columns before placing formwork	Yes / No			Yes / No		
b	Wet formwork and steel before pouring concrete	Yes / No			Yes / No		
c	Use concrete within 90 minutes of mixing with water if from factory	Yes / No			Yes / No		
d	If mixed manually use in less than 30 minutes	Yes / No			Yes / No		
e	Ensure toothed areas of columns completely filled with concrete	Yes / No			Yes / No		
f	Use rod to consolidate concrete around reinforcement, then hammer on formwork	Yes / No			Yes / No		
g	Pour column in one day to the same height as wall	Yes / No			Yes / No		
h	Pour columns after minimum 1m wall built	Yes / No			Yes / No		
i	Complete entire beam within one day	Yes / No			Yes / No		
j	Have plastic on standby, cover if it rains	Yes / No			Yes / No		
k	Roughen upper surface of plinth beam concrete with cross lines made with a trowel	Yes / No			Yes / No		
l	Cover freshly cast concrete with a tarp or empty cement bags	Yes / No			Yes / No		
m	Cure for minimum 1 week by sprinkling clean water twice a day	Yes / No			Yes / No		
9	**CONCRETE INSPECTION**	**Planned?**	**Date**	**Comments**	**Built?**	**Date**	**Comments**
a	If steel showing, demolish and rebuild	Yes / No			Yes / No		
b	Remove the border of slab and/or beams after 48 hours	Yes / No			Yes / No		
c	Any cracks larger than 3 mm	Yes / No			Yes / No		
d	Many cracks in one location	Yes / No			Yes / No		
e	Diagonal or vertical cracks anywhere in the beam	Yes / No			Yes / No		
f	*If any of the above exist, demolish concrete and repour*	Yes / No			Yes / No		

Homeowner Signature: _____

Engineer Signature: _____

Builder Signature: _____

FOUNDATION CHECKLIST

Homeowner: _____

Designer: _____   Address: _____

Builder: _____

		Planned?	Date	Comments	Built?	Date	Comments
1	**SITE LAYOUT**						
a	Excavation consistent with plan	Yes / No			Yes / No		
b	Excavation level	Yes / No			Yes / No		
c	Batterboard completed	Yes / No			Yes / No		
d	Excavation lines at right angles	Yes / No / NA			Yes / No / NA		
2	**FOUNDATION EXCAVATION DEPTH**	**Planned?**	**Date**	**Comments**	**Built?**	**Date**	**Comments**
a	Depth of foundation excavation 80 cm (minimum)	Yes / No			Yes / No		
b	Depth of excavation in natural ground 50 cm (min)	Yes / No			Yes / No		
3	**FOUNDATION MINIMUM WIDTH**	**Planned?**	**Date**	**Comments**	**Built?**	**Date**	**Comments**
a	Hard soil (rock and compact gravel) = 40 cm	Yes / No			Yes / No		
b	Clay soil or clay sand = 50 cm	Yes / No			Yes / No		
c	Sandy soils = 70 cm	Yes / No			Yes / No		
4	**FOUNDATION TRENCH EXCAVATION**	**Planned?**	**Date**	**Comments**	**Built?**	**Date**	**Comments**
a	Remove water from excavation	Yes / No / NA			Yes / No / NA		
b	Remove loose soil from excavation	Yes / No			Yes / No		
c	Remove any organic debris or tree trunks	Yes / No / NA			Yes / No / NA		
d	Bottom flat and level	Yes / No			Yes / No		
e	Bottom flat and level	Yes / No			Yes / No		
5	**FOOTING BASE LAYER**	**Planned?**	**Date**	**Comments**	**Built?**	**Date**	**Comments**
a	Minimum 5 cm thickness	Yes / No			Yes / No		
b	Mix 1:10 (lean concrete)	Yes / No			Yes / No		
6	**STONE MASONRY CONTINUOUS FOOTING**	**Planned?**	**Date**	**Comments**	**Built?**	**Date**	**Comments**
a	Use cut, angular stones	Yes / No			Yes / No		
b	Use mix 1:5 for mortar	Yes / No			Yes / No		
c	Fill all gaps between stones with mortar	Yes / No			Yes / No		
d	Use cross stones every 1 m	Yes / No			Yes / No		
e	Scarify top for good contact	Yes / No			Yes / No		
f	Cure properly	Yes / No			Yes / No		
g	Backfill with compacted soil in 10 cm lifts	Yes / No			Yes / No		

7	TIE COLUMN ANCHORS	Planned?	Date	Comments	Built?	Date	Comments
a	Use four 10 mm bars at each tie column location with 6 mm stirrup cage	Yes / No			Yes / No		
b	Bend bottom of 10 mm bars in four directions to create self supporting rebar cage	Yes / No			Yes / No		
c	Minimum 5 cm void around rebar cage	Yes / No			Yes / No		
d	Use 3 cm concrete spacers to achieve proper concrete cover below bars	Yes / No			Yes / No		
8	UTILITY PIPES	Planned?	Date	Comments	Built?	Date	Comments
a	Do not put piping through plinth beam	Yes / No			Yes / No		
b	Put piping through strip footing	Yes / No			Yes / No		
c	Void around pipe filled by bigger pipe	Yes / No			Yes / No		

Homeowner Signature: _____

Engineer Signature: _____

Builder Signature: _____

MASONRY WALL CHECKLIST

Homeowner: _____

Designer: _____ Address: _____

Builder: _____

1	MORTAR MIXING	Planned?	Date	Comments	Built?	Date	Comments
a	Use mortar 1:3 mix	Yes / No			Yes / No		
b	Use clean, fine river sand	Yes / No			Yes / No		
c	Use clean water (not salty or muddy)	Yes / No			Yes / No		
d	Use Portland Cement	Yes / No			Yes / No		
e	Mix on a clean, concrete or asphalt surface, not on dirt	Yes / No			Yes / No		
f	Use a mechanical mixer if possible	Yes / No			Yes / No		
g	Batch out gravel, then sand, then cement	Yes / No			Yes / No		
h	Mix gravel, sand, and cement dry, then add water	Yes / No			Yes / No		
i	Turn over 3 times or until color is uniform	Yes / No			Yes / No		
j	Do not use too much water! Add water slowly	Yes / No			Yes / No		

2	WALL MASONRY	Planned?	Date	Comments	Built?	Date	Comments
a	Wet concrete blocks prior to use	Yes / No			Yes / No		
b	Use a line and deadman	Yes / No			Yes / No		
c	Prop up column steel so it remains plumb	Yes / No			Yes / No		
d	Overlap blocks 1/3 of their length, stack 1 run at a time (not diagonally)	Yes / No			Yes / No		
e	When setting block, vibrate block or tap block with a trowel	Yes / No			Yes / No		
f	Tooth wall at tie columns and openings by 1/3 block	Yes / No			Yes / No		
g	Maintain minimum 3.5 cm between block and column tie	Yes / No			Yes / No		
h	Joint thicknesses 10 - 15 mm	Yes / No			Yes / No		
i	Prepare a reasonable amount of mortar to avoid adding water	Yes / No			Yes / No		
j	Check the wall is plumb	Yes / No			Yes / No		
k	Wet the wall 3 times per day for 3 days	Yes / No			Yes / No		
l	Check the top of the wall is level	Yes / No			Yes / No		

3	ELECTRICAL and PLUMBING	Planned?	Date	Comments	Built?	Date	Comments
a	Never break the wall to place electrical or plumbing	Yes / No / NA			Yes / No / NA		
b	Leave free space for utility piping	Yes / No / NA			Yes / No / NA		
c	Fill voids around conduit hidden in the wall with mortar or concrete	Yes / No / NA			Yes / No / NA		

Homeowner Signature: _____

Engineer Signature: _____

Builder Signature: _____

FLOOR AND ROOF CONCRETE SLAB CHECKLIST

Homeowner: _____ Address: _____

Designer: _____

Builder: _____

1 SLAB LAYOUT	Planned?	Date	Comments	Built?	Date	Comments
a	Minimum thickness = 1/20 of the span	Yes / No			Yes / No		
b	*FORMWORK:*						
1	Posts are straight, vertical, braced and placed on level ground	Yes / No			Yes / No		
2	Horizontal boards are good quality and straight edged	Yes / No			Yes / No		
c	*REINFORCEMENT:*						
1	Primary and secondary reinforcement size and spacing per details	Yes / No			Yes / No		
2	Bars hooked at ends of slab	Yes / No			Yes / No		
3	Upper reinforcement supported by 'chairs'	Yes / No			Yes / No		

2 PAN AND JOIST (LIGHTWEIGHT) SLABS	Planned?	Date	Comments	Built?	Date	Comments
a	Concrete blocks, bricks, or polystyrene formwork	Yes / No			Yes / No		
b	Primary reinforcement placed between forms centered in space	Yes / No			Yes / No		
c	Secondary reinforcement placed on top of forms with 3 cm spacers	Yes / No			Yes / No		
d	Primary and secondary reinforcement size and spacing per details	Yes / No			Yes / No		
e	Primary and secondary reinforcement hooked into ring beams	Yes / No			Yes / No		
f	Pipes in slab do not cross joists unless double joists are used	Yes / No / NA			Yes / No / NA		

3 CONCRETE POURING and CURING	Planned?	Date	Comments	Built?	Date	Comments
a	Verify that water and drainage pipes do not leak	Yes / No / NA			Yes / No / NA		
b	Wet formwork and pans before pouring concrete	Yes / No			Yes / No		
c	Use concrete within 90 minutes of mixing with water if from factory	Yes / No			Yes / No		
d	If mixed manually use in less than 30 minutes	Yes / No			Yes / No		
e	Use needle vibrator or rod to consolidate concrete around reinforcement	Yes / No			Yes / No		
f	Use straightedge to smooth and level the concrete	Yes / No / NA			Yes / No / NA		
g	Pour slab in one day	Yes / No			Yes / No		
h	Cure with little ponds on the slab filled with water for at least 2 weeks	Yes / No			Yes / No		
i	Remove supporting formwork for 3 weeks	Yes / No			Yes / No		
j	Cover vertical rebar projecting above the slab with lean concrete	Yes / No / NA			Yes / No / NA		

Homeowner Signature: _____

Engineer Signature: _____

Builder Signature: _____

Appendix

A. TABLE OF CONCRETE, MORTAR AND PLASTER MIXES

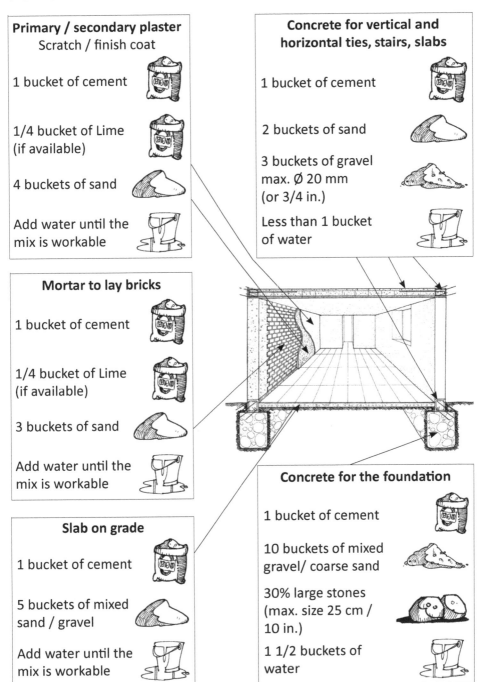

Primary / secondary plaster
Scratch / finish coat

1 bucket of cement

1/4 bucket of Lime
(if available)

4 buckets of sand

Add water until the
mix is workable

Mortar to lay bricks

1 bucket of cement

1/4 bucket of Lime
(if available)

3 buckets of sand

Add water until the
mix is workable

Slab on grade

1 bucket of cement

5 buckets of mixed
sand / gravel

Add water until the
mix is workable

**Concrete for vertical and
horizontal ties, stairs, slabs**

1 bucket of cement

2 buckets of sand

3 buckets of gravel
max. Ø 20 mm
(or 3/4 in.)

Less than 1 bucket
of water

Concrete for the foundation

1 bucket of cement

10 buckets of mixed
gravel/ coarse sand

30% large stones
(max. size 25 cm /
10 in.)

1 1/2 buckets of
water

Figure 236: The various mixes to use in the construction process

161

B. HOW TO DETERMINE THE NUMBER OF SHEAR WALLS

The earthquake resistance of your house depends on the right amount of shear walls in each direction and their correct position and size. The quantity of shear walls depends on the type of soil. With the aid of an example house we will calculate the correct number of shear walls.

1. Identify the level of seismic hazard in your area. Ask your municipal engineer for this information. We consider three levels of seismic hazard:

    • Low: Peak ground acceleration (PGA) lower than 0.08 g

    • Moderate: PGA between 0.08 g and 0.25 g

    • High: PGA between 0.25 g and 0.4 g

2. Determine the type of soil on which your building will stand. We consider three types of soil:

    • Rock and firm soil (A)

    • Compact granular soil (B)

    • Soft clay soil or soft sand (C)

3. Determine the percentage of shear walls needed in each direction with the help of the following tables. The amount of shear walls depends on the type of masonry units used: solid bricks or blocks allow for less shear walls than hollow bricks or hollow concrete blocks.

Table for **solid** clay bricks or solid concrete blocks

	Hazard level Low (below 0.08 g)	Hazard level moderate (0.08 - 0.25 g)		Hazard level High (0.25 - 0.4 g)	
	Soil type A + B + C	Soil type A	Soil type B + C	Soil type A	Soil type B + C
Ground floor	1.5 %	1.5 %	2.0 %	3.0 %	4.5 %
Upper floor	1.0 %	1.0 %	1.0 %	1.5 %	2.5 %

Table for **hollow** clay bricks or hollow concrete blocks

	Hazard level Low (below 0.08 g)	Hazard level moderate (0.08 - 0.25 g)		Hazard level High (0.25 - 0.4 g)	
	Soil type A + B + C	Soil type A	Soil type B + C	Soil type A	Soil type B + C
Ground floor	1.5 %	1.5 %	3.5 %	4.0 %	6.5 %
Upper floor	1.0 %	1.0 %	2.0 %	2.0 %	3.5 %

Figure 237: Percentage of shear walls of a building

4. Calculate the ceiling area of each storey.

5. Calculate the required footprint (horizontal area) of shear walls for each floor, in each direction, by multiplying the ceiling area with the percentage identified in the tables above. In other words, for a two storey building:

    • Multiply the ceiling area of the ground floor with the shear wall percentage of the ground floor.
      This will give you the amount of shear walls needed at ground floor level.

    • Then multiply the ceiling area of the upper floor with the shear wall percentage of the upper floor. This will give you the amount of shear walls needed at the upper floor.

    For a one-storey building use the percentages indicated for the ground floor. This will ensure that the amount of shear walls will be correct when you decide in the future to add an upper floor to your house.

6. Calculate the necessary shear wall lengths for each direction and each floor. Divide the result for the ground floor by the thickness of your shear walls. Do the same for the upper floor.

7. Make sure that the length of each independent confined shear wall element is at least two thirds of its height (from below plinth to above ring beam / slab).

8. Verify that the total length of all shear wall elements of one façade is more than half the total length of that façade.

9. Make sure that the shear walls are distributed evenly. Symmetrical layouts are best.

Example

To illustrate the process, let's take the example of the following house to be built with 20 cm hollow concrete blocks.

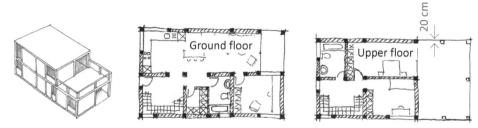

Figure 238: Example house for the calculation of shear walls

1. Seismic hazard zone: High

2. Type of soil: A = Rock or firm soil

3. Shear wall percentage for hollow concrete blocks: at ground floor = 4 %, at upper floor = 2 %

4. Calculate the ceiling area above each floor:

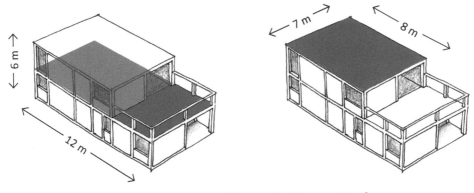

Ceiling above ground floor: 7 x 12 m = 84 m²

Ceiling above upper floor: 7 x 8 m = 56 m²

Figure 239: Calculate the ceiling area of each storey

5. Calculate the shear wall footprint by multiplying the ceiling area of each storey with the corresponding shear wall percentage:
   - Ground floor: 84 m² x 0.04 = 3.36 m² of shear wall footprint in each direction.
   - Upper floor: 56 m² x 0.02 = 1.12 m² of shear wall foot print in each direction.

6. Calculate the necessary shear wall lengths for each direction and each floor:
   - Groud floor: 3.36 m² / 0.2 m = 16.8 m. You need a total of 16.8 m of shear walls in each direction.
   - Upper floor: 1.12 m² / 0.2 m = 5.6 m. You need a total of 5.6 m of shear walls in each direction.

7. Determine the minimum length of independent shear wall elements: Storey height 3 m x 2/3 = 2 m minimum length of shear wall elements which can be taken into consideration.

Total length of shear wall footprint: 7.8 + 2.6 + 2.2 + 4.8 + 2.2 = 19.4 m;
19.4 m > 16.8 m = **OK**

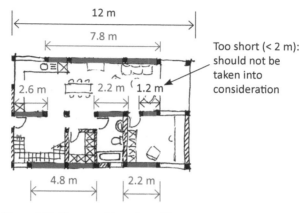

Too short (< 2 m): should not be taken into consideration

Figure 240: Shear wall length verification

8. Verify the ratio shear wall length / total façade length.
   - Façade 1: 7.8 m / 12 m = > 50 % = **OK**
   - Façade 2: 7 m / 12 m = > 50 % = **OK**

9. The shear walls are distributed evenly.

Repeat the last two operations for the shear walls in the other direction.

7. Total length of shear wall footprint: 2.5 + 1.9 + 2.5 + 3.6 + 3.6 = 14.1 m;
   14.1 m < 16.8 m = **NOT OK**

Use solid concrete blocks which need a shear wall percentage of 3 % instead of 4 % (or add 2.7 m of wall).

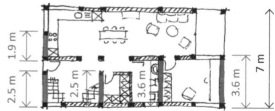

   - Ground floor:
     84 m² x 0.03 = 2.52 m² of shear wall footprint in **this** direction.
   - Ground floor: 2.52 m² / 0.2 m =12.6 m. You need a total of 12.6 m of shear walls in **this** direction.

Total length of shear wall footprint: 2.5 + 1.9 + 2.5 + 3.6 + 3.6 = 14.1 m;
14.1 m > 12.6 m = **OK**

8. Verify the ratio shear wall length / total façade length.
   - Façade 3: 4.4 m / 7 m = > 50 % = **OK**
   - Façade 4: 3.6 m / 7 m = > 50 % = **OK**

Repeat the operations for the shear walls of the upper floor.

BIBLIOGRAPHY AND RECOMMENDED READING

Confined Masonry Network:

- *Codes and Standards*, http://www.confinedmasonry.org/existing-codes-and-standards

- *Construction Guidelines*, http://www.confinedmasonry.org/existing-construction-guidelines

AFPS (2004): *Construction parasismique des maisons individuelles aux Antilles "Guide CP-MI Antilles"*, Association Française du Génie Parasismique, Paris, France

AIS (2001): *Manual de construcción, evaluacción y rehabilitación sismo resistente en viviendas de mampostería*, Asociacción Colombiana de Ingegnería Sísmica, Colombia

Blondet M. et al. (2005): *Construction and Maintenance of Masonry Houses, for Masons and Craftsmen*, Pontificia Universidad Católica del Perú and SENCICO, Lima, Peru

Boen T. and associates (2009): *Constructing Seismic Resistant Masonry Houses*, World Seismic Safety Initaitive (WSSI) and United Nations Center for Regional Development (UNCRD), Disaster Management Planning Hyogo Office, Japan / Indonesia

Brzev S (2007): *Earthquake Resistant Confined Masonry Construction*, NICEE, Indian Institute of Technology Kanpur, India

Earthquake Hazard Centre, *Newsletter*, Victoria University of Wellington, New Zealand

ERRA (2008): *Compliance Catalogue, Guidelines for the Construction of Compliant Rural Houses*, Earthquake Reconstruction and Rehabilitation Authority, Pakistan

Arya A et al. (2013): *Guidelines for Earthquake Resistant Non-Engineered Construction*, UNESCO, Paris, France

Meli R., Brzev S. et al. (2011): *Seismic Design Guide for Low-Rise Confined Masonry Buildings*, Confined Masonry Network, World Housing Encyclopedia, EERI and IAEE, Oakland CA, USA

MTPTC (2010); *Guide de bonnes pratiques pour la construction de petits bâtiments en maçonnerie chaînée en Haïti*, Ministère des Travaux Publiques, Transports et Communications (MTPTC) and Ministère de l'Intérieur et des Collectivités Territoriales (MICT), Haiti

Murty C.V.R. et al. (2006): *At Risk: The Seismic Performance of RC Frame Buildings with Masonry Infill Walls*, World Housing Encyclopedia, EERI and IAEE, Oakland CA, USA

Murty C.V.R. (2002-4): *Earthquake Tips 1 - 24*, Indian Institute of Technology Kanpur (IITK) and Building Materials and Technology Promotion Council (BMTPC), India

NSET (2005): *Earthquake Resistant Construction of Buildings, Curriculum for Mason Training, Guidelines for Training Instructors*, Asiam Disaster Preparedness Center (ADPC) and National Society for Earthquake Technology (NSET), Nepal

Schacher T (2009): *Confined Masonry for One and Two Storey Buidlings in Low-Tech Environments, A guidebook for technicians and artisans*, NICEE, Indian Institute of Technology Kanpur, India

Schacher T. (2011): *La Maçonnerie Chaînée, Cours de formation pour maçon et formateurs (13 Powerpoint lessons)*, Centre de Compétence Reconstruction, Swiss Agency for Development and Cooperation SDC, Port au Prince, Haiti

SDC (2013): *Gid pou konstwi kay pi solid (Guidebook to build safer houses in confined masonry)*, Centre de Compétence Reconstruction, Swiss Agency for Development and Cooperation, Port au Prince, Haiti

SDC (2017): *Guía para la construcción de viviendas sismo-resistentes en mampostería confinada*, (revised and updated version of Haitian manual) Swiss Agency for Development and Cooperation SDC, Quito, Ecuador

Virdi K., Rashkoff R.: *Confined Masonry Construction*, City University London, UK

Milton Keynes UK
Ingram Content Group UK Ltd.
UKHW051156210924
1771UKWH00012B/58